Cambridge Elements ☰

Elements of Paleontology
edited by
Colin D. Sumrall
University of Tennessee

MOLECULAR PALEOBIOLOGY OF THE ECHINODERM SKELETON

Jeffrey R. Thompson
The University of Southampton

CAMBRIDGE
UNIVERSITY PRESS

Shaftesbury Road, Cambridge CB2 8EA, United Kingdom

One Liberty Plaza, 20th Floor, New York, NY 10006, USA

477 Williamstown Road, Port Melbourne, VIC 3207, Australia

314–321, 3rd Floor, Plot 3, Splendor Forum, Jasola District Centre,
New Delhi – 110025, India

103 Penang Road, #05–06/07, Visioncrest Commercial, Singapore 238467

Cambridge University Press is part of Cambridge University Press & Assessment,
a department of the University of Cambridge.

We share the University's mission to contribute to society through the pursuit of
education, learning and research at the highest international levels of excellence.

www.cambridge.org
Information on this title: www.cambridge.org/9781009179751

DOI: 10.1017/9781009179768

First published 2022

A catalogue record for this publication is available from the British Library.

ISBN 978-1-009-17975-1 Paperback
ISSN 2517-780X (online)
ISSN 2517-7796 (print)

Molecular Paleobiology of the Echinoderm Skeleton

Elements of Paleontology

DOI: 10.1017/9781009179768
First published online: November 2022

Jeffrey R. Thompson
The University of Southampton

Author for correspondence: Jeffrey R. Thompson, j.r.thompson@soton.ac.uk

Abstract: The echinoderms are an ideal group to understand evolution from a holistic, interdisciplinary framework. The genetic regulatory networks underpinning development in echinoderms are some of the best known for any model group. Additionally, the echinoderms have an excellent fossil record, elucidating in detail the evolutionary changes underpinning morphological evolution. In this Element, the echinoderms are discussed as a model group for molecular palaeobiological studies, integrating what is known of their development, genomes, and fossil record. Together, these insights shed light on the molecular and morphological evolution underpinning the vast biodiversity of echinoderms, and the animal kingdom more generally.

Keywords: skeletogenesis, development, evo-devo, sea urchins, gene regulatory networks

ISBNs: 9781009179751 (PB), 9781009179768 (OC)
ISSNs: 2517-780X (online), 2517-7796 (print)

Contents

1 Introduction

Molecular paleobiology provides a promising avenue to merge data from deep-time, molecular biology and genomics, gaining insights into the evolutionary process at multiple levels (Peterson et al., 2007; Wörheide et al., 2016). Given the vast literature concerning gene expression and developmental evolution in echinoderms (Davidson et al., 2002a; Revilla-i-Domingo et al., 2007; Oliveri et al., 2008; Dylus et al., 2018; Shashikant et al., 2018), the echinoderm skeleton is a model for molecular paleobiological studies (Bottjer et al., 2006). I will begin with an overview of molecular paleobiology and the skeletogenic process in echinoderms, as well as a discussion of what gene regulatory networks (GRNs) are, and why they are of interest to paleobiologists. I then highlight recent advances in the evolution of the echinoderm skeleton from both paleobiological and molecular/functional genomic perspectives, highlighting examples where diverse approaches provide complementary insight. Finally, I discuss the enormous potential of molecular paleobiology of echinoderms as a growing and fruitful field of research.

1.1 Molecular Paleobiology

The fossil record provides invaluable insight into evolutionary transitions in morphology through deep time. Morphological evolution is the result of changes in organismal development, and thus a holistic understanding of morphological evolution is not complete without an understanding of the evolution of the molecular and genomic mechanisms whose evolution underlie changes in *animal development*. While the fossil record provides the most direct evidence of the history of life, the fossil record cannot provide direct insight into the molecular or genomic mechanisms which operate during development. Likewise, direct interrogation of *genomes*, *gene* function and regulation, and the molecular mechanisms operating during development is limited to extant taxa, or at least relatively geologically recent ones. Molecular paleobiology seeks to use the tools of molecular biology and genomics, including gene expression, functional genomics, molecular phylogenetics, and divergence time estimation, to address questions primarily formulated from paleobiology (Peterson et al., 2007; Wörheide et al., 2016). Molecular paleobiological approaches have shed light on diverse paleontological and evolutionary bio-logical questions, from the origin and interrelationships of paleobiologically important animal groups (Sperling et al., 2011; Vinther et al., 2012) to dating key transitions in the history of life (Schirrmeister et al., 2015; Lozano-Fernandez et al., 2016; Fleming et al., 2018; Howard et al., 2020). Furthermore, data from the fossil record are able to provide insight into extinct and transitional morphologies and body plans not seen among extant lineages and can be used to generate testable hypotheses about morphological character evolution, ontogenetic transitions, and homology (Shubin et al., 1997; Shubin et al., 2009; Garwood et al., 2014; Tweedt, 2017; Chipman and Edgecombe, 2019). The beauty of molecular palaeobiology is this two-way transfer of information. Fossils can inform on experiments to be carried out at the bench, and DNA sequence data can be used to understand the timing of evolutionary events and mechanisms of evolutionary change.

2 The Echinoderm Skeleton in Development and Evolution

The echinoderm skeleton provides a fantastic system with which to understand the evolution of a key morphological innovation from multiple perspectives spanning across disciplines. From a palaeontological perspective, the $CaCO_3$ skeleton of echinoderms has provided most members of the phylum Echinodermata with an exceptional fossil record (Figure 1). This has facilitated their use to understand morphological, paleoecological, and trait-based evolution across long and deep timescales (Syverson and Baumiller, 2014; Hopkins and Smith, 2015; Wright, 2017; Cole et al., 2019; Bauer, 2021; Clark et al., 2020; Deline et al., 2020; Mongiardino Koch and Thompson, 2021). In addition to this rich insight into paleobiological questions, the echinoderm skeleton is also a cutting-edge model system used to understand how the regulation, function, and expression of genes directs the processes of animal development and evolution (Davidson et al., 2002a; Revilla-i-Domingo et al., 2007; Oliveri et al., 2008; Dylus et al., 2018; Shashikant et al., 2018).

With the publication of the sequenced genome of the purple sea urchin, *Strongylocentrotus purpuratus*, in 2006 (Sodergren et al., 2006), understanding the regulatory genomic mechanisms by which the echinoderm skeleton develops became more tractable than ever before. Analyses of the structure, arrangement, and number of genes and families of genes involved in skeletogenesis (the molecular and developmental processes that build the skeleton) laid the groundwork for the current, meticulous understanding of the genetic regulatory mechanisms which orchestrate development of the larval skeleton in sea urchins (Livingston et al., 2006; Oliveri et al., 2008; Ettensohn, 2009; Sharma and Ettensohn, 2010; Rafiq et al., 2012; Rafiq et al., 2014; Shashikant et al., 2018). In addition to providing insight into how the regulatory genome directs the development of the larval skeleton, the publication of the sea urchin genome also allowed for novel and creative insight into the evolution of the echinoderm skeleton in deep time (Bottjer et al., 2006). With their manuscript entitled "Paleogenomics of Echinoderms," which was published alongside the

Figure 1 Examples of the skeletons of adult echinoderms from the crown group classes.

The biomineralized echinoderm skeleton is comprised of numerous $CaCO_3$ plates, which are present in fossil members of each of the extant classes. (1) shows the skeleton of the stem group echinoid *Pholodechinus brauni*. The test of stem group echinoids is made up of numerous columns of both ambulacral and interambulacral plates, most of which bore spines. (2) shows the stem group crinoid *Griphocrinus pirovanoi*. The crown of the animal includes the many calcified plates of the calyx, in addition to multiple arm plates. (3) shows the biomineralized skeleton of the ophiocistioid stem group holothurian *Eucladia*

S. purpuratus genome in a special issue of *Science*, Bottjer et al. (2006) made one of the first explicit attempts to link the evolution of genes to the initial appearance of a morphological feature, in this case the echinoderm skeleton, in the fossil record. A multitude of new comparative data from genomes and *transcriptomes* (all expressed genes in an organism or tissue determined through RNA sequencing) of other, nonechinoid, echinoderms have come to light in the years since the publication of "Paleogenomics of Echinoderms." Nevertheless, there is little doubt that "Paleogenomics of Echinoderms" remains an influential paper in molecular, genomic, and paleobiological studies of the echinoderm skeleton and remains an important contribution highlighting multidisciplinary approaches to understand the evolution of a major evolutionary innovation underpinning the origin of a diverse phylum, the Echinodermata.

The goal of this contribution is to highlight a number of novel insights into the deep-time evolution of the echinoderm skeleton that have taken place since the publication of "Paleogenomics of Echinoderms" in 2006. To do this, I will first discuss the skeletons of echinoderms and the cells that are responsible for building them. Next, I will take some space to explain the conceptual and practical relevance of GRNs, and their various components, in development. Thereafter, what is known about the genetic regulatory networks that build the skeletons of larval and adult echinoderms will be reviewed. I will then focus on work using deep-time perspectives to understand the evolution of the echinoderm skeleton, focusing on examples from both the fossil record and extant organisms. Finally, I would like to touch on a number of areas where I feel that future integration of paleontological data and data from analyses of gene expression and gene function can synergistically inform the evolution of the echinoderm skeleton and the evolution of animal diversity more generally.

2.1 Echinoderm Skeletons

The activities of GRNs result in the development and growth of animal morphology, like the echinoderm skeleton. As skeletons make up the bulk

---

Caption for Figure 1 (cont.)

johnsoni. Unlike crown group holothurians, this stem group member had a plated test consisting of numerous imbricating calcified plates, as well as plated tube feet and a central calcified jaw apparatus not unlike the Aristotle's lantern of echinoids. (4) shows the skeleton of the fossil ophiuroid *Palaeocoma milleri*, Courtesy of Tim Ewin. (5) shows the fossil asteroid *Alkaidia sumralli*, with the many-plated morphology of the sea star skeleton, Courtesy of T. Ewin and A. Gale.

of the animal fossil record, GRNs are the blueprints for much of the data preserved therein. The exceptional fossil record of echinoderms is due to the abundance of their adult skeletons in the rock record. This CaCO₃ skeleton consists of numerous skeletal plates, or ossicles, which can be either abutting, imbricating (Smith, 1980), or sutured together with interlocking struts (Smith, 1990; Grun and Nebelsick, 2018). Much of the adult skeleton is embedded within the dermis, and is *mesodermal* in origin. It is the structure and arrangement of the skeleton that gives extant echinoderms their characteristic fivefold symmetry and is responsible for the characteristic morphology of most classes (Figure 1). The skeleton of most adult echinoderms is characterized by a distinct trabecular microstructure, termed *stereom*, which contains an occluded *protein* matrix and sits surrounded within organic tissues termed the stroma (see Gorzelak (2021), this volume). Echinoderm skeletons can be highly variable in terms of their morphology, and this diversity of echinoderm skeletons forms the basis for much of the morphological diversity discussed in this volume.

In addition to their disparate adult skeletons, there are a diversity of larval skeletons found across the Echinodermata (Figure 2A) (Mortensen, 1921). These have a miniscule fossil record (Reich, 2021; Deflandre-Rigaud, 1946), but they have formed the basis for a number of important studies attempting to infer larval ecology (Strathmann, 1971, 1975; Pennington and Strathmann, 1990), the effects of changes in ocean chemistry on larval mortality and vitality (Brennand et al., 2010; Byrne et al., 2011; Byrne et al., 2013), and attempts to infer echinoderm phylogeny (Wray, 1992; Strathmann and Eernisse, 1994; Smith, 1997). The larval skeleton of echinoids, known as the echinopluteus, is an extensive structure consisting of between four and six skeletal elements that assist in orientation while in the water column (Figure 2A, C–D) (Pennington and Strathmann, 1990). The echinopluteus is characteristically referred to as easel-shaped, and most echinoplutei have pairs of elongate, ciliate larval arms, which protrude from the body around the mouth (Mortensen, 1921). The skeletal elements of echinoplutei can be rod-like or fenestrated, having the appearance of small ladders (Mortensen, 1921; Wray, 1992). Whereas some direct developing echinoids have lost the echinopluteus larvae throughout the course of evolution, the presence of an echinopluteus has been demonstrated to be the ancestral state, at least among crown group echinoids (Wray, 1992).

Extensive larval skeletons are not unique to the echinoids. Many indirect developing ophiuroids have planktonic ophioplutei larvae, which also have an elongate larval skeleton with up to four pairs of skeletal arms (Figure 2A) (Mortensen, 1921; Raff and Byrne, 2006; Gliznutsa and Dautov, 2011).

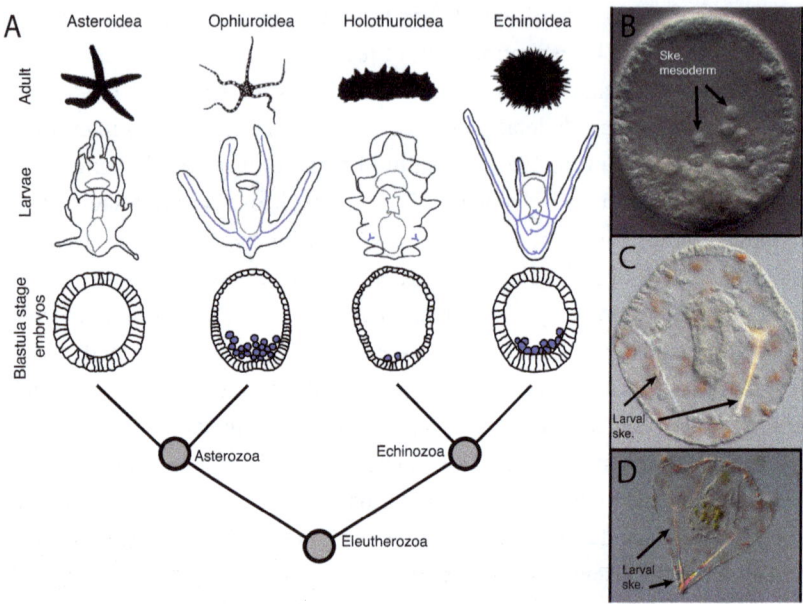

Figure 2 Eleutherozoan echinoderm larval skeletons. (A) shows the phylogenetic distribution of eleutherozoan echinoderm larvae and blastula stage embryos with skeletal mesodermal cells. Larval skeletons and skeletal mesodermal cells are shown in blue. Extensive larval skeletons are present in the ophiopluteus larvae of ophiuroids and the echinopluteus larvae of echinoids. Auricularia larvae of holothurians have a miniscule skeleton consisting of two small spicules present in posterior end of the larvae. The bipinnaria larvae of asteroids lack a larval skeleton, and a mesodermally derived larval skeletogenic cell lineage. (B) shows skeletogenic mesodermal cells in mesenchyme blastula stage embryo of the echinoid *S. purpuratus*. (C) shows larval skeleton in prism stage embryo of *S. purpuratus*. There are two bilaterally arranged skeletal elements that are derived from triradiate spicules. (D) shows larval skeleton in the pluteus larvae of *S. purpuratus*. All larvae are from indirect developing species. No known crinoid larvae are indirect developers, thus crinoids are excluded from the diagram. Ske; skeleton.

Though not supported by recent phylogenomic analyses (Cannon et al., 2014; Telford et al., 2014; Mongiardino Koch et al., 2018), the gross morphological similarity between the ophiopluteus and echinopluteus had previously been taken as evidence for a close phylogenetic relationship of echinoids and ophiur-oids – the so-called cryptosyringid hypothesis (Smith, 1997). Within the ophiur-oids, the adult body plan develops either from a rudiment in the ophiopluteus or through a distinct additional metamorphic stage known as the vittelaria (Mortensen, 1921; Byrne and Selvakumaraswamy, 2002). The ancestral condition within ophiuroids is still not well-known, as the developmental mode is not well-known for most ophiuroids; however, reduced ophiopluteal skeletal elements in vittelaria suggest ophioplutei are the ancestral condition (Byrne and Selvakumaraswamy, 2002). Furthermore, though these different larval types are distributed broadly across ophiuroid phylogeny (McEdward and Miner, 2001; O'Hara et al., 2014), the morphological similarity of ophioplutei to echinoplutei also suggests that the ophiopluteus may be the ancestral condition within ophiuroids (Raff and Byrne, 2006), and analyses of the echinoderm skeletogenic cell suggest a single evolutionary origin for the *cell type* that builds both echinoid and ophiuroid larval skeletons (Erkenbrack and Thompson, 2019).

The holothurians, or sea cucumbers, lack the extensive elongate skeletons present in echinoids, but they too have larval skeletons (Figure 2A). In contrast to echinoids and ophiuroids, direct development is the most common developmental mode among holothurians (Sewell and McEuen, 2002). Indirect developing holothurian larvae are referred to as auricularia, and in many auricularia, small spicules are found along the posterior end of the larvae (Woodland, 1907b; McCauley et al., 2012). Auricularia larvae of the synaptid holothurians are also known to possess skeletal wheels-like ossicles, similar to the wheel-shaped ossicles present in adult holothruians and are the ontogenetic outcomes of the aforementioned spicules (Stricker, 1985, 1986; Woodland, 1907b). The larval skeleton of holothurians has often been overlooked relative to that of ophiuroids and echinoids, and many accounts of the holothurian larval skeleton predate the advent of scanning electron microscopy (Woodland, 1907b; Mortensen, 1921). While this is the case, understanding the growth and development of the holothurian larval skeleton has a crucial role to play in understanding the evolution and origin of echinoderm skeletons (McCauley et al., 2012; Erkenbrack and Thompson, 2019).

Asteroids are unique among the *eleutherozoans,* in that their larvae lack a larval skeleton (Figure 2A). There exists a diversity of larvae among the asteroids, whose evolutionary histories provide an ideal model system to study the evolution of larval morphology (Hart et al., 1997; Carter et al., 2021). However, because none of these larvae have skeletons, they will not be discussed in detail here. Though some larval asteroids have elongated larval arms, which may serve analogous purposes to those of echinoplutei and ophioplutei, these larval arms have no endoskeleton supporting them (Mortensen, 1921). In addition to the absence of a larval skeleton, asteroid larvae also lack skeletogenic cells, which give rise to the skeleton in the larvae of other eleutherozoan classes. It is this absence of a skeleton in asteroids that has resulted in much of the controversy surrounding the homology or homoplasy of the echinoid or holothurian larval skeletons. All known crinoid larvae are direct developing, and as such lack a larval skeleton (McEdward and Miner, 2001). Thus, they also do not factor into this discussion of larval skeletons.

2.2 Skeletogenic Cells in Echinoderms

The echinoderm skeleton is secreted by a distinct population of cells, known as skeletogenic cells, or *sclerocytes*. Nascent biomineral is deposited in a *vacuole* within a cellular *syncytium* made up of the fusion of the cell processes and cytoplasm (Figure 3) (Märkel et al., 1986; Smith, 1990; Gliznutsa and Dautov, 2011). Both larval and adult echinoderms possess skeletons, although much more is known about the molecular signature of larval skeletogenic cells in echinoderms than those that secrete the adult skeleton (Killian and Wilt, 2008). In particular, one population of embryonic/larval skeletogenic mesodermal cells of echinoids (Figure 3C–E), known alternatively as primary *mesenchyme cells* or PMCs, are exceptionally well characterized from a molecular perspective (Oliveri et al., 2008; Rafiq et al., 2012; Barsi et al., 2014; Rafiq et al., 2014; Barsi et al., 2015). In all known eleutherozoans, these embryonic and larval skeletogenic cells are mesenchymal, meaning in this case that they are loosely associated and mobile. Skeletogenic cells are first specified early in development, prior to *blastula stage embryos*, and go on to build the larval skeleton in indirect developing echinoids, ophiuroids, and holothurians (Okazaki, 1975; Gliznutsa and Dautov, 2011; McCauley et al., 2012; Dylus et al., 2016). After their specification, the presumptive skeletogenic cells undergo an *epithelial* to mesenchymal transition, migrating from the wall of blastula or *gastrula* stage embryos, or the ingressing *archenteron* (dependent upon the taxon) into the *blastocoel* during development (Figures 2B, 3C–D) (Wray and McClay, 1988; Wu and McClay, 2007; Wu et al., 2007). Once in the blastocoel, they form a syncytium, in which the $CaCO_3$ of the larval spicule and skeleton is deposited (Figures 2C–D, 3E) (Woodland, 1907b; Smith, 1990; Gliznutsa and Dautov, 2011). Though the homology of these cells across echinoderm classes is debated (Erkenbrack and Thompson, 2019), the cellular movements and processes associated with larval skeletogenic cells are found across echinoderms with larval skeletons and characterize larval skeletogenesis in echinoids, ophiuroids, and holothurians.

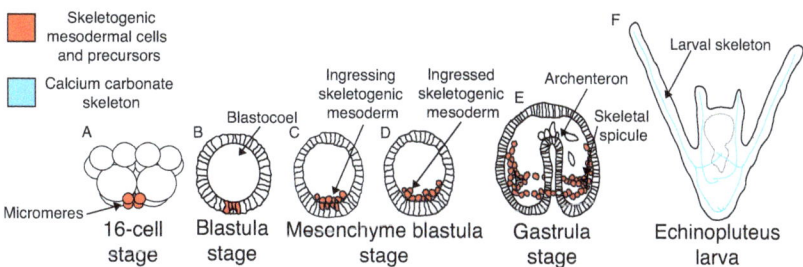

Figure 3 Schematic diagram showing select developmental stages of an indirect developing echinoderm larva. This diagram is based on that of a camarodont echinoid, and is, in many ways, representative of other skeleton-bearing echinoderm larvae. (A) shows the sixteen-cell stage, during which, in camarodont echinoids and most other known euechinoid echinoids, the cell lineage which will become the skeletogenic cells is first specified. The skeletogenic cells are derived from the four micromeres (shown in red), which are present at the vegetal pole (bottom) of the embryo. Later during development, (B) shows the blastula stage embryo that begins at the approximately 128 cell stage in the camarodont *S. purpuratus*. The blastula consists of a sphere of cells surrounding an open cavity, called the blastocoel. In blastula stage embryos, the cells that will give rise to the skeletogenic cells, and which have descended from the micromeres, are located in the vegetal pole of the embryo as part of the epithelium of the blastula wall (shown in red). During the mesenchyme blastula stage (C), the skeletogenic cells migrate from the epithelial wall of the blastula into the blastocoel. This migration is known as an epithelial to mesenchymal transition, as the skeletogenic cells (red), which were part of the epithelium of the blastula, are now loose and mobile. All of the skeletogenic cells migrate into the blastocoel (D), where they will later secrete the biomineralized tri-radiate spicules of the skeleton. Later during development, (E) shows a gastrula stage embryo, at which point the archenteron, which will eventually attach to the wall of the ectoderm to form the gut, has formed from an invagination in the vegetal pole of the embryo (the blastopore). During gastrulation, the skeletogenic cells (red) have arranged themselves into bilaterally symmetrical ventro-lateral clusters on either side of the embryo and begun to secrete the tri-radiate spicules that will grow to form the larval skeleton (blue). (F) shows a simplified echinopluteus larvae. The skeletogenic cells are not shown, but the larval skeleton, which now comprises four elongate skeletal elements, can be seen in blue. Some differences, such as the presence of four distinct micromeres, do exist between camarodont echinoids and other echinoderms, such as cidaroid echinoids, which have a variable number of micromeres, and ophiuroids and holothurians, which lack any micromeres at all. Furthermore, as shown in Figure 2, there are also differences in the morphologies

Less is known about the molecular and cellular mechanisms operating in the skeletogenic cells of adult echinoderms, though their morphology and function during skeletogenesis has been characterized in a number of echinoderm groups. Adult echinoderm skeletogenic cells have been most thoroughly characterized in echinoids, where skeletal cells have been classified into at least two distinct groups: the sclerocytes and the odontoblasts (Märkel et al., 1986; Märkel et al., 1989). Additionally, sclerocytes have been identified and characterized in regenerating ophiuroid arms (Piovani et al., 2021). In echinoids, sclerocytes are the skeleton-secreting cells that secrete biomineral throughout the majority of the animal, and odontoblasts are those skeletogenic cells that are responsible for biomineralization of the continuously growing teeth of the Aristotle's lantern (Märkel et al., 1986; Märkel et al., 1989). Much of the adult echinoderm skeleton, including echinoid test plates and spines, asterozoan arm plates, and ossicles of the holothurian body wall, lie embedded within the dermis, and thus further discussion will focus on the sclerocytes. In growing echinoderm skeletal elements, such as the margins of interambulacral plates of the echinoid test, there is a higher density of skeletal cells in areas of active skeletogenesis (Shimizu, 1997). Like the skeletogenic cells of the larvae, sclerocytes are responsible for biomineralization within cytoplasmic sheaths or, when multiple cells have merged, a syncytial vacuole (Heatfield and Travis, 1975; Märkel et al., 1986, 1989; Stricker, 1986; Dubois and Jangoux, 1990; Smith, 1990; Ameye et al., 1999; Piovani et al., 2021). Within the cytoplasmic sheath of this syncytial vacuole, the skeleton is surrounded by a matrix coat, comprised of polysaccharides and proteins (Märkel et al., 1989; Ameye et al., 1999). These sclerocytes have characteristic outgrowths that contact the stereom (Dubois and Jangoux, 1990; Ameye et al., 1999). During regeneration, the sheath of these skeletal cells surrounds the ends of growing stereom trabeculae, and new biomineral is deposited within the vacuole of the sheath (Heatfield and Travis, 1975; Dubois and Jangoux, 1990; Ameye et al., 1999).

Echinoderm skeletogenic cells were first identified and characterized based upon their function and morphology, although most recent work has begun to understand these cells in the context of the suites of genes they express (Rafiq et al., 2014; Piovani et al., 2021), and the role of these genes in building the skeleton. The expression and activity of these genes is the product of the

---

Caption for Figure 3 (cont.)

of the larvae and skeletons of other echinoderms. Diagrams in B–D are modified from Erkenbrack and Thompson (2019) and E is modified from McClay et al. (2020) with permission from Elsevier.

molecular interactions encoded in the genome and characterized by the skele-togenic GRN (Oliveri et al., 2008; Shashikant et al., 2018).

3 What Is a Developmental Gene Regulatory Network?

In recent years (though see Valentine and Campbell, 1975), there has been interest from the palaeontological community in *developmental gene regulatory networks*, largely because of their potential explanatory power in understanding the evolution of animal body plans (Davidson and Erwin, 2006; Erwin and Davidson, 2009; Thompson et al., 2015, 2017; Bottjer, 2017; Erwin, 2020) and homology of morphological characters (Wagner, 2007). Because of this interest, and because of the crucial role GRNs play in understanding the evolution and development of the echinoderm skeleton (Dylus et al., 2016; Thompson et al., 2017; Dylus et al., 2018; Shashikant et al., 2018), it is worth taking some space to explain exactly what GRNs are, and why they are relevant in development and evolution.

Developmental GRNs describe the interactions of genes and gene products (e.g proteins) during the course of organismal development (Figure 4) (Davidson et al., 2002a; Peter and Davidson, 2015). Gene regulation is the process by which a gene, or a gene product, regulates the expression of another gene (Figure 4A–C). Genes encoding for proteins can regulate the expression of other genes both positively (Figure 4A; upregulation) and negatively (Figure 4C; downregulation), resulting in the fluctuating spatial and temporal patterns of gene expression across various cell and tissue types during development (Wray and Lowe, 2000; Levine and Davidson, 2005; Davidson and Levine, 2008). GRNs consist of regulatory inter-actions of hundreds of different genes and proteins, all of which interact directly or indirectly to ensure development proceeds correctly (Barsi et al., 2015; Peter and Davidson, 2015; Peter and Davidson, 2017; Khor et al., 2019). The structure of a GRN reflects the timing of gene expression and regulation, with more upstream components expressed earlier in development than those that are more down-stream. This results in a hierarchical structure observed in many GRNs (Figure 4A–C), which allows for their constituent parts, or subcircuits, to be broken down and thought of as individual modules (Levine and Davidson, 2005; Davidson and Erwin, 2006; Peter and Davidson, 2016; Peter and Davidson, 2017). GRNs are comprised of multiple different types of genes, including those that encode *transcription factors*, *signaling molecules*, and *differentiation proteins* (Sections 3.1, 3.2, and 3.3), all of which have different functions in development.

3.1 Transcription Factors: Regulating Gene Expression

Transcription factors are proteins, encoded for by genes, that regulate the expres-sion of other genes through the physical process of binding to DNA sequences

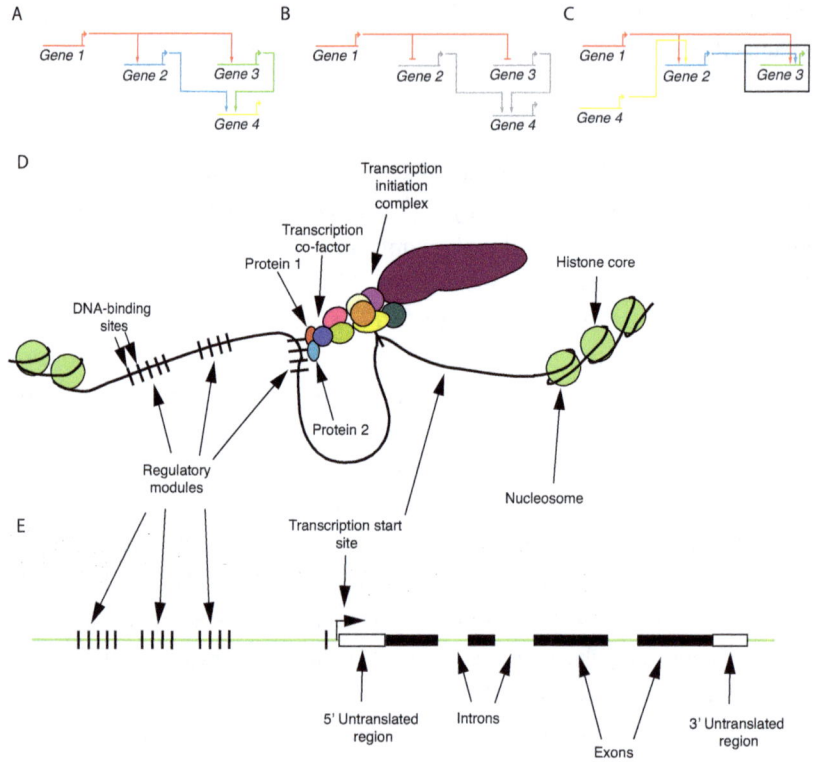

Figure 4 Diagrammatic representation of gene regulatory networks (GRNs). A–C show wiring diagrams of simple GRNs. In both (A) and (B), there are four genes: genes 1, 2, 3, and 4. The arrows indicate physical regulatory interactions, where a protein encoded for by each gene binds to DNA in a regulatory region of a downstream target gene. In (A), gene regulation is positive, where the binding of gene 1 to its downstream targets 2 and 3 results in an increase in their expression (upregulation). Genes 2 and 3 also positively regulate their downstream target, gene 4. In (B), gene 1 acts as a repressor, and results in the repression of its downstream targets 2 and 3, and in turn their target gene 4. In (C), genes 1 and 4 both regulate gene 2, and gene 2 and gene 1 both regulate gene 3. This scenario is more representative of biological reality, as numerous transcription factors interact combinatorially to regulate gene expression during animal development. (D) shows a simplified diagram of the cis-regulatory region of gene 3 from (C) with the binding of the transcription factors encoded for by genes A and B binding to DNA-binding sites in their respective regulatory modules upstream of (before) the transcription initiation complex and the transcription start site. (E) shows the organization of gene 3. The cis-regulatory region, as shown in (D), is located upstream of the transcription start site and the transcribed portions of the gene. Exons are transcribed but introns

around the genes they regulate (Figure 4D–E) (Wray et al., 2003; Gilbert, 2006; Peter and Davidson, 2015). These sequences, called DNA binding sites, can be located in the noncoding regions flanking the *exons* where the *transcriptional machinery* (including the *RNA polymerase* and other aspects of the *transcription initiation complex*) binds, called the *promoter* (Watson et al., 2008). Additionally, DNA binding sites can be located in regulatory modules, consisting of numerous individual binding sites, that can be located thousands of nucleotides from the promoter, including in gene *introns*. These regulatory modules located far from the transcriptional start sites are called *enhancers* (Figure 4D–E).

When a transcription factor binds to a regulatory sequence of a gene, it interacts with proteins involved in RNA polymerase binding, *histone* modifying proteins, and other transcription factors to stabilize the transcription initiation complex, open *nucleosomes,* or otherwise facilitate or repress transcription of the target gene (Wray et al., 2003; Gilbert, 2006; Watson et al., 2008) (Figure 4D). Transcription factors are thus responsible for modulating both the amount and timing of expression of the genes that they regulate. They are responsible for both positive regulation, where their activity results in higher levels of expression of their downstream target genes, and negative regulation or repression, where their binding to a regulatory region on their downstream target results in a reduction or silencing of its expression (Revilla-i-Domingo et al., 2007; Watson et al., 2008; Peter and Davidson, 2015).

Transcription factors and *transcription co-factors* act in combination to regulate the process of gene expression during development (Figure 4D–E) (Wray et al., 2003; Davidson, 2006; Gilbert, 2006; Watson et al., 2008). Because transcription factors are responsible for gene regulation, they are among the most important components of a GRN (Davidson et al., 2002a; Davidson, 2006; Peter and Davidson, 2015). The sum of all transcription factors expressed in a particular cell or set of cells is called the regulatory state, and different regulatory states are responsible for the existence of differentially specified cell fates during development (Peter and Davidson, 2015). In this way, transcription factors are crucial components underlying the specification

---

Caption for Figure 4 (cont.)

are not. The 5' and 3' untranslated regions (so-named because of their position relative to the gene) will not be translated into the protein but are transcribed.
(E) and (D) are simplified from Wray et al. (2003) with permissions from Oxford University Press, to which I refer the reader for a more in-depth discussion of transcriptional regulation.

of the diverse cell types present throughout the course of animal development and evolution (Arendt, 2008; Arendt et al., 2016). Transcription factors are also among the more upstream members of GRNs, and can regulate many hundreds of downstream targets, including signaling molecules and differentiation genes (Rafiq et al., 2012; Rafiq et al., 2014; Khor et al., 2019).

3.2 Signaling Molecules: Communication between Cells during Development

Cell-to-cell signaling provides a means by which different tissues and organs are able to communicate during development. Cell-to-cell signaling typically involves a specific *ligand* sending the signal and binding to a specific receptor molecule on the surface of the cell that receives the signal. Signaling can take place through direct contact between two cells, through the diffusion of the ligand across small distances in the developing animal, or over long distances via fluid, such as blood (Gilbert, 2006). The receptor protein spans the cell membrane, with an extracellular portion outside the cell, a transmembrane region, and a cytoplasmic region inside of the cell. The extracellular portion receives a signal from other cells by binding to the ligand. This results in a change of shape of the receptor protein on the inside and outside of the cell, which in turn results in a series of enzymatic changes inside the cell, typically ending in the activation of a transcription factor (Gilbert, 2006). This series of interactions is called a *signal transduction cascade.*

During development, these signaling pathways operate between different cell and tissue types, and often provide spatial cues that result in the positioning of developing morphological structures, or the *induction* of new organs and tissue types (Gilbert, 2006; Duloquin et al., 2007). Signaling molecules play an important role in GRNs directing development, as they are responsible for activating particular GRN subcircuits in spatially distinct suites of cells during growth and development (Peter and Davidson, 2015; Peter and Davidson, 2017). Signaling molecules act as bridges during development, allowing for the output of a GRN in one cell to activate more downstream components in adjacent or nearby cells. Because of their role in cell-to-cell communication, signaling molecules help to confer modularity and hierarchy to the activity of GRNs (Levine and Davidson, 2005).

3.3 Differentiation Gene Batteries: The Interface with Morphology

Among the most downstream components of a GRN are the differentiation genes (Davidson et al., 2002a; Davidson et al., 2002b). Differentiation genes do not regulate the expression of other genes, and are so-called because they are crucial players in the processes of *cellular differentiation*, the process by which cells change into more specific cell types through the course of development. These differentiation genes and the proteins for which they encode are often expressed in specific cell types, including skeletal cells, photoreceptors, or neurons (Wilt et al., 2008; Barsi et al., 2014; Barsi et al., 2015; Garner et al., 2016). Differentiation proteins are expressed at the most peripheral portions of a GRN (Davidson et al., 2002a), building morphological structures at a molecular level and encoding for proteins found in distinct tissue types such as skeletal tissue and nerves, and structures that are responsible for pigmentation and color. Because of their position at the periphery of a network, and because they do not regulate other genes, they have also been proposed to undergo the highest rates of evolution within a GRN (Davidson and Erwin, 2006; Erwin and Davidson, 2009; Peter and Davidson, 2011). Because differentiation genes are not responsible for the regulation of other genes, they are more likely to be able to undergo evolutionary changes due to mutation without catastrophic consequences on developmental processes. This may also explain why differentiation genes often appear to have undergone high rates of gene duplication throughout the history of life (Livingston et al., 2006).

4 The Echinoderm Skeletogenic Gene Regulatory Network

The GRN directing development of the embryonic and larval skeleton of echinoderms is one of the best-known developmental GRNs in animal development (Figure 5). Though it has primarily been studied in echinoids, echinoderm larval skeletogenic cells express a distinct set of transcription factors, which interact together with signaling molecules and differentiation genes to build the larval skeleton. It is thus crucial to note that most of the activity of the echinoderm skeletogenic GRN takes place specifically in the skeletogenic cells that build the skeleton (Figures 2B, 3). It would be impossible to discuss the role and function of all genes comprising the skeletogenic GRN in this contribution, so I herein choose to focus on some of the best characterized, and potentially most important, genes in the network. An important note moving forward concerns the nomenclature of genes and proteins. If the name of a molecule is shown in italics, it refers to the DNA sequence or *mRNA* transcript of the gene, for example, *Alx1*. If the name is shown without italics, for instance, Alx1, the name refers to the protein that has been translated from the mRNA transcript. Because transcription factors are proteins, when discussed in the context of their regulatory function, typically the unitalicized version is used. If discussing the gene that encodes for the transcription factor, as is often the case when discussing gene expression, the name is italicized.

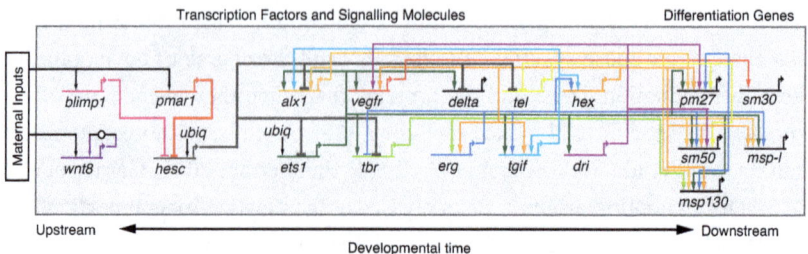

Figure 5 Simplified skeletogenic GRN from embryonic and larval *Strongylocentrotus purpuratus*. To the left are more upstream transcription factors and signaling molecules such as *Pmar1* and *HesC*,which are involved in skeletogenic cell specification and operate earlier in development. Toward the center of the diagram are transcription factors such as *Alx1, Ets1/2,* and *Tbr,* all of which are crucial components responsible for conferring skeletogenic cell identity in larval *S. purpuratus*, and which regulate the expression of up to hundreds of differentiation genes involved in skeletal growth and biomineral deposition. Also toward the center is the signaling molecule *VegfR*, which has a crucial role in positioning of the larval spicule during skeletogenesis. Additional genes, such as *Hex, Dri, Erg,* and *Tel* are transcription factors with roles in skeletogenesis and regulate the expression of downstream differentiation genes. At the right are differentiation genes such as the spicule matrix genes *SM30, SM50,* and *MSP130,* whose expression is necessary for normal biomineral deposition and growth. Black arrow depicts more upstream or downstream components of the GRN. Arrows in wiring diagram indicate positive regulatory interactions, while plungers represent repressive regulatory interactions. Components of the double-negative gate (DNG) have been shown depicted with bold lines.

4.1 Transcription Factors in the Echinoderm Skeletogenic Gene Regulatory Network

The most upstream components of the skeletogenic GRN are regulatory genes that encode for transcription factors (Figure 5), and one of the most extensively studied is the transcription factor *Alx1* (Ettensohn et al., 2003; Khor and Ettensohn, 2020). Alx1 has regulatory inputs into more than 400 other genes that are differentially expressed in *S. purpuratus* skeletogenic cells (Rafiq et al., 2014; Khor et al., 2019), and *Alx1* expression has been identified in the skeletogenic cells of all larval echinoderms (with skeletogenic cells) thus far examined (Ettensohn et al., 2003; Yamazaki et al., 2010; McCauley et al., 2012; Yamazaki et al., 2014; Erkenbrack and Davidson, 2015; Yamazaki and

Minokawa, 2015; Dylus et al., 2016; Koga et al., 2016; Khor and Ettensohn, 2017; Morgulis et al., 2019). Additionally, *Alx1* is expressed in sites of skeletogenesis in adult or post-metamorphic echinoids (Gao and Davidson, 2008; Gao et al., 2015), asteroids (Koga et al., 2014; Koga et al., 2016), and ophiuroids (Czarkwiani et al., 2013; Piovani et al., 2021).

In addition to Alx1, several other transcription factors have been implicated as crucial components of the skeletogenic GRN (Figure 5). The gene *Ets1/2* is expressed in the skeletogenic and non-skeletogenic mesodermal cells of embryonic and larval eleutherozoans (Rizzo et al., 2006). *Ets1/2* also has regulatory inputs into hundreds of genes expressed in skeletogenic cells (Rafiq et al., 2014) in *S. purpuratus,* and *knockdown* (reduction or blocking of gene expression) of this gene in multiple larval echinoderms results in a failure of mesodermal cells to properly differentiate, and thus failed skeletogenesis (Kurokawa et al., 1999; Koga et al., 2010). *Tbr,* a transcription factor gene expressed in skeletogenic cells and broader mesodermal tissues across embryonic and larval echinoderms, is also necessary for larval skeletogenesis in *S. purpuratus* (Croce et al., 2001; Oliveri et al., 2008; McCauley et al., 2012; Yamazaki et al., 2014; Erkenbrack and Davidson, 2015; Yamazaki and Minokawa, 2015; Dylus et al., 2016). In contrast to *Alx1*, knockdown of *Tbr* in cidaroid echinoids does not result in a reduction of skeletogenic cells (Yamazaki et al., 2014). Furthermore, *Tbr* is not expressed in skeletogenic centers in adult echinoids or ophiuroids (Gao and Davidson, 2008; Czarkwiani et al., 2013), indicating it may be a later evolutionary addition to the *S. purpuratus* larval skeletogenic GRN (Erkenbrack and Thompson, 2019).

There are numerous other transcription factors from the *S. purpuratus* skeletogenic GRN that are expressed in the skeletogenic cells of other echinoderms, including Erg, Hex, TGIF, and Jun (Figure 5). The spatial expression of these other transcription factors has not been surveyed at the same taxonomic breadth as for *Alx1, Ets1/2,* or *Tbr,* and thus the demonstrated extent of their involvement in skeletogenesis at broad taxonomic scales is lesser known. All of these genes have been shown to be expressed in the skeletogenic cells of *S. purpuartus*, however, and some have demonstrated expression in skeletogenic cells of other echinoids, ophiuroids, and holothurians (McCauley et al., 2012; Russo et al., 2014; Dylus et al., 2016; Erkenbrack et al., 2016; Piovani et al., 2021). Although the spatial expression of these transcription factors has not been widely surveyed from a phylogenetic standpoint, their functional importance in the *S. purpuratus* skeletogenic GRN has been validated (Oliveri et al., 2008), and thus their role in skeletogenesis outside of *S. purpuratus* remains a fruitful avenue of research for understanding the evolution of gene function in GRNs.

4.2 Signaling Molecules in the Echinoderm Skeletogenic Gene Regulatory Network

Aside from transcription factors, signaling molecules play an important role in the skeletogenic GRN for echinoderms (Figure 5). Though numerous signaling pathways are involved in regulation and development of the echinoderm skeleton (Adomako-Ankomah and Ettensohn, 2014), I will focus on the two that have been the most extensively studied in echinoderms: the vascular endothelial growth factor (VEGF) and fibroblast growth factor (FGF) signaling pathways. These two pathways are necessary components of normal skeletogenesis in echinoderms. The VEGF pathway is involved in vascularization and blood vessel growth and development in vertebrates. In echinoderms, however, it underlies skeletogenesis (Morgulis et al., 2019). In both larval and adult skeletogenesis, the signaling ligand, *Vegf3,* is expressed in *ectodermal* tissues overlying the skeletogenic mesodermal cells (Duloquin et al., 2007; Morino et al., 2012; Adomako-Ankomah and Ettensohn, 2013; Adomako-Ankomah and Ettensohn, 2014; Erkenbrack and Petsios, 2017; Czarkwiani et al., 2021; Morgulis et al., 2019). *VegfR-10-Ig,* which receives this signal from the overlying *Vegf3*-expressing ectodermal cells, is expressed specifically in these skeletogenic cells. Knockdown of *Vegf3* or *VegfR-10-Ig* results in reduced formation of skeletal biomineral and expression of skeletogenic genes, as well as incorrect spatial positioning of the skeletogenic mesodermal cells (Duloquin et al., 2007). This suggests that *Vegf3* sends spatial positioning cues to the skeletogenic mesodermal cells and that *VegfR-10-Ig* expression in the skeletogenic cells regulates the expression of downstream biomineralization genes involved in skeletogenesis (Duloquin et al., 2007; Morgulis et al., 2019). This pattern, with *VegfR-10-Ig* expressed in the skeletogenic mesoderm and *Vegf3* expressed in the overlying ectoderm, has been identified in larval regular euechinoid echinoids (Duloquin et al., 2007; Adomako-Ankomah and Ettensohn, 2013; Morgulis et al., 2019), cidaroids (Erkenbrack and Petsios, 2017), ophiuroids (Morino et al., 2012; Czarkwiani et al., 2021) as well as in skeletogenesis in adult or metamorphic asteroids (Morino et al., 2012; Koga et al., 2014), ophiuroids (Czarkwiani et al., 2021), and in the case of only *VegfR-10-Ig,* echinoids (Gao and Davidson, 2008; Gao et al., 2015). Also, importantly, neither *Vegf3* nor *VegfR-10-Ig* were identified in early development of sea stars, which lack a larval skeletogenic cell (Morino et al., 2012). Taken together, this evidence implicates the role of VEGF signaling in the spatial positioning of the skeleton, and further regulation of skeletogenic genes, thus making it a crucial component of the skeletogenic GRN in echinoderms.

In addition to VEGF signaling, the FGF pathway has also been identified as an important part of the skeletogenic GRN. Like VEGF, the gene encoding the FGF ligand, *Fgf9/16/20* (herein referred to as *Fgf*), is expressed in the ectoderm of larval sea urchins, though in largely distinct territories from the expression of *Vegf3* (Röttinger et al., 2008; Adomako-Ankomah and Ettensohn, 2013). In contrast to *Vegf3*, however, *Fgf* is expressed not only in the ectoderm but also in the skeletogenic mesodermal cells themselves. The FGF signaling receptor, *FgfR2*, is expressed specifically in the skeletogenic mesodermal cells, as is the case with *VegfR*, and other components of the skeletogenic GRN (Röttinger et al., 2008). Inhibition of FGF signaling results in disrupted skeletogenesis and the downregulation of the biomineralization genes SM30 and SM50 (Röttinger et al., 2008). Both VEGF and FGF signaling are thought to underlie the branching, anastamozing morphology of stereom. When larval spicules are cultured from isolated skeletogenic mesodermal cells, and thus in the absence of ectodermal signals, the triradiate spicules lack any branching (Okazaki, 1975). Furthermore, in regenerating asteroid spines, stereom at the spine margin, adjacent to the ectoderm, grows longitudinally via branching, suggesting that signals from the ectoderm modulate the direction and morphology of stereom growth (Dubois and Jangoux, 1990). Given that VEGF signaling is involved in patterning the anastomosing tubular morphology of blood vessels in vertebrates, it seems likely that it may fill a similar role in patterning the tubular, branching morphology of the echinoderm skeleton (Morgulis et al., 2019).

4.3 Differentiation Gene Batteries in the Echinoderm Skeletogenic Gene Regulatory Network

The most downstream components of the echinoderm skeletogenic GRN are the differentiation genes (Figure 5), which include biomineralization genes, as well as genes that assist in ion transport and morphogenetic processes like cell fusion (Shashikant et al., 2018). These biomineralization genes are responsible for depositing nascent $CaCO_3$ and building the biomineral structure and are thus very much the building blocks for the echinoderm biomineral skeleton. Though there are hundreds of downstream differentiation genes involved in echinoderm skeletogenesis, the most well-known of these biomineralization genes are those of the spicule matrix (SM) and MSP130 families (Figure 5). The SM genes, which include *SM30*, S*M37*, *SM50, C-lectin*, and *PM27,* encode for c-lectin-type *extracellular matrix* proteins and are expressed during sea urchin biomineral growth and (at least some of which) are occluded in the skeletal organic matrix (Ameye et al., 1999; Livingston et al., 2006; Mann et al., 2008a, 2008b, 2010). Crucially, the SM family of genes are specific to echinoids and appear to be absent from the genomes and transcriptomes of other echinoderms (Zhang et al., 2017; Dylus et al., 2018). The SM genes are expressed in the skeletogenic cells of embryonic and larval echinoids (Guss and Ettensohn, 1997b), and the proteins they encode for are occluded within the organic matrix of the larval skeleton (Mann et al., 2010). Knockdown of some SM genes in the larvae results in a failure of larval skeletal elements to form or elongate, though just how important particular SM genes are varies on a protein-by-protein basis (Wilt et al., 2008; Wilt et al., 2013). In addition to their role in larval skeletogenesis, SM30 and SM50 proteins have been identified specifically in the skeletogenic cells and occluded skeletal matrix of adult echinoid skeletal tissues (Ameye et al., 1999; Thompson et al., 2021). Though the SM genes are not present in nonechinoid echinoderms, it seems likely that there are analogous, yet nonhomologous genes that fill a similar role in other echinoderm groups such as ophiuroids (Czarkwiani et al., 2021).

In addition to the SM genes, the MSP130 family of genes are a well-known component of echinoderm skeletogenesis (Figure 5). MSP130 is a cell-surface *glycoprotein* expressed in skeletogenic cells and skeletal tissues of echinoderms across classes and life history stages (Anstrom et al., 1987; Leaf et al., 1987; Guss and Ettensohn, 1997a; Minokawa et al., 1997; Livingston et al., 2006; Mann et al., 2008a, 2008b, 2010; Chiaramonte et al., 2020). Homologues of MSP130 genes have been identified across numerous biomineralizing metazoans (Marie et al., 2011; Cameron and Bishop, 2012; Ettensohn, 2014; Szabó and Ferrier, 2015), as well as in the genomes, transcriptomes, and *proteomes* of crinoids, echinoids, ophiuroids, and holothurians (Livingston et al., 2006; Zhang et al., 2017; Dylus et al., 2018; Davidson et al., 2020; Li et al., 2020). Similarly to the SM genes, the MSP130 genes have undergone extensive *gene duplication* throughout their evolutionary history, with numerous *paralogues* and closely related genes in echinoids (Livingston et al., 2006; Ettensohn, 2014; Dylus et al., 2018; Davidson et al., 2020). Noteworthy, however, is the fact that phylogenetic analyses support independent, homoplastic duplications of MSP130 and related genes in echinoids and ophiuroids, underpinning the crucial role of this gene in echinoderm biomineralization (Dylus et al., 2018).

Though they have been less extensively studied than the SM and MSP130 genes, additional differentiation genes include those that encode for extracellular matrix proteins or those with roles in ion transport, cell fusion, and biomineralization (Rafiq et al., 2012). A selection of these are *CAN* and *Caral7* that encode for carbonic anhydrases, a family of enzymes that catalyze the conversion of CO_2 and H_2O to H^+ and HCO_3^- ions for skeletogenesis and pH regulation (Chow and Benson, 1979; Mitsunaga et al., 1986; Livingston et al., 2006;

Mann et al., 2008b), the adhesion protein, KirrelL, which is crucial for *filo-podial fusion* of skeletogenic cells (Ettensohn and Dey, 2017), and P16 and P19, proteins with a poorly characterized function that are crucial for sea urchin larval skeleton elongation and have been implicated in skeletogenesis across echinoderms (Cheers and Ettensohn, 2005; Costa et al., 2012; Dylus et al., 2018). It is the downstream differentiation genes that are responsible for the process of *morphogenesis*, and in this way, they provide a direct link between the regulatory transcription factors and signaling molecules, and animal morphology.

Though there remains more work to be done in characterizing the suites of genes expressed in echinoderm skeletogenic cells at wide phylogenetic scales, what has been done so far shows that the GRN contains hundreds of genes, many with shared and distinct functions (Shashikant et al., 2018). How the suite of expressed genes and their functions evolve provides novel insight into how morphologies are likely to have evolved in both shallow and deep time.

5 Evolution of the Echinoderm Skeleton

5.1 What Can We Learn from the Fossil Record?

The exceptional echinoderm fossil record provides unparalleled insights into phenotypic evolution, providing clues as to the potential operation of GRNs in deep time. Though it is impossible to know with certainty the genomic regulatory networks and patterns of gene expression that were present in extinct taxa, the fossil record can provide insight into the phenotypic patterns of evolution that are inherently the morphological outcomes of the activity of gene expression and regulation. This in turn can provide insight into possible molecular scenarios underlying morphological evolution in deep time, and the fundamental evolutionary patterns used to generate hypotheses concerning GRN evolution.

5.1.1 Macroevolutionary Trends in the Evolution of Echinoderm Body Plans

A recent example where the fossil record of echinoderms has been used to understand the macroevolutionary consequences of GRNs concerns the work of Deline et al. (2020) (Figure 6). GRNs have an inherently hierarchical structure, which has been proposed to underlie the differential evolvability of morpho-logical characters whose development they direct (Davidson and Erwin, 2006; Erwin and Davidson, 2009; Peter and Davidson, 2015; Peter and Davidson, 2016; Peter and Davidson, 2017). Davidson and Erwin (2006) hypothesized that the hierarchical position of genes and GRN subcircuits within a developmental GRN may have corresponded to the morphological characters whose

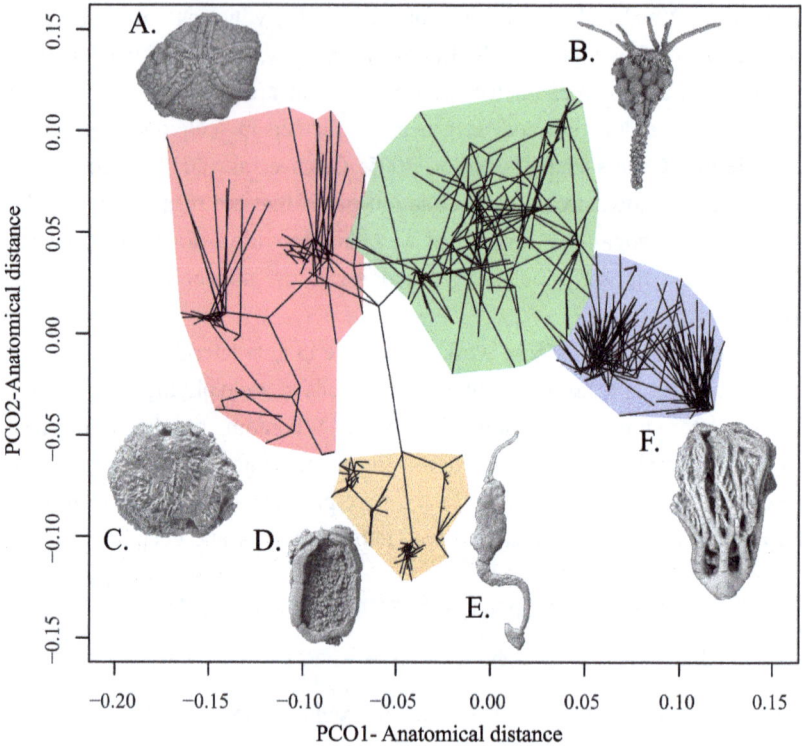

Figure 6 Phylomorphospace showing the distribution of Cambrian and Ordovician echinoderms based on their morphological disparity in an analysis from Deline et al. (2020). The morphospace shows two axes from a principal coordinate analysis based on Gower's similarity metric. Lines represent the phylogenetic relationships of all taxa included in the morphospace. Analyses of morphological disparity resulted in four main echinoderm body plans (highlighted in color). Red shows the radial attached/mobile body plan, green the stalked radial blastozoan body plan, blue the stalked radial crinoid body plan, and yellow the non-radial body plan. The characters used to create this morphospace show no hierarchical signal in evolution, casting doubt on the relationship between character burden and the genetic regulatory underpinning of morphological characters. (A) is the Ordovician edrioasteroid *Edrioaster*. (B) is the Cambrian eocrinoid *Sinoeocrinus*. (C) is the Ordovician echinoid *Bromidechinus*. (D) is the Cambrian ctenocystoid *Ctenocystis*. (E) is the Cambrian solute *Coleicarpus*. (F) is the Ordovician crinoid *Anomalocrinus*. Courtesy of Brad Deline.

development they direct. They hypothesized that downstream differentiation genes at the periphery of a GRN are responsible for the evolution of species-level characters, whereas the tightly and recursively wired subcircuits of transcription factors expressed earlier in development control phenotypic characters manifested at the phylum and class levels, such as symmetry, or the presence or absence of limbs. This hypothesis is an extension of the work of Riedl (1977), who proposed the concept of differential and hierarchical evolvability of phenotypic characters underlain by a concept he termed "burden." Reidl's "burden" reflects the hierarchically arranged interdependence of organismal characters, and was invoked as an explanation for why some characters, those that define organismal body plans, were conserved across large animal groups, while others, those with less burden, appear to have higher rates of phenotypic evolution (Riedl, 1977; Schoch, 2010).

Using a large matrix of morphological characters from Cambrian and Ordovician echinoderms, Deline et al. (2020) analyzed morphological disparity of the echinoderm skeleton (Figure 6) and used estimates of phylogenetic signal to evaluate patterns of morphological evolution during the initial burst of echinoderm morphological diversification. To assess the relationship between character burden and evolvability, characters were coded based on the number of morphological characters contingent upon their presence or absence in the character matrix. This value was then compared to the phylogenetic signal of each character, that is, the phylogenetically based tendency for closely related taxa to have more similar traits to each other than to taxa they are less related to (Pagel, 1999; Borges et al., 2019). Their analyses revealed no clear relationship between phylogenetic signal and number of contingent characters, indicating that the phylogenetic distribution of characters does not seem to be directly related to the burden, or hierarchical rank of characters. This result countered the predicted model of Riedl (1977), and the extension to GRN theory proposed by Davidson and Erwin (2006), suggesting that characters with a high burden, which are thought to be conserved and relatively impervious to evolutionary change, are in fact not so. This implies not only that these body plan level characters are more evolvable than expected, but also that the hierarchical nature of the GRNs directing their development do not have a directly hierarchical effect on the evolution of morphology.

5.1.2 Reduction of the Holothurian Skeleton

Extant holothurians are characterized by a highly reduced skeleton relative to other echinoderms (Figure 7) (Smith and Reich, 2013). Most other extant and fossil echinoderms have a robust calcium carbonate skeleton, but the skeleton of most crown group holothurians is comprised primarily of microscopic calcium carbonate spicules embedded in their body wall (Stricker, 1985; Stricker, 1986; Woodland, 1906; Woodland, 1907a, 1907b). The morphological transitions

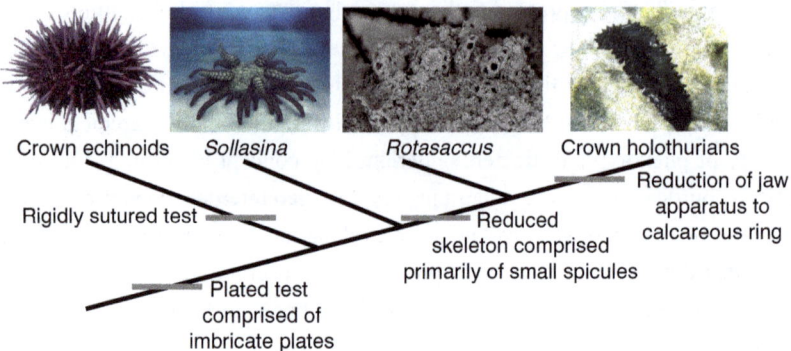

Crown echinoids   *Sollasina*   *Rotasaccus*   Crown holothurians

Rigidly sutured test   Reduction of jaw
                                        Reduced   apparatus to
                                        skeleton comprised   calcareous ring
                                        primarily of small spicules
                    Plated test
                    comprised of
                    imbricate plates

Figure 7 Phylogenetic tree showing the reduction of the adult skeleton in holothurians throughout the course of echinozoan evolution. Extant and extinct echinoids have a test comprised of multiple robust $CaCO_3$ plates, which became rigidly sutured near the transition from the echinoid stem group to crown group (Thompson et al., 2020). Most stem group echinoids, however, had tests comprised of multiple imbricate plates. Like stem group echinoids, stem group holothurians like the ophiucistioid *Sollasina* also had skeletons comprised of multiple imbricate plates and calcified jaw apparatuses (Rahman et al., 2019). Along the lineage leading to crown group holothurians, however, a reduction of the skeleton took place to tiny, ossicles embedded in the dermis. This is first present in ophiocistioids like *Rotasaccus* (which have jaw apparatuses), and even more extensively realized in crown group holothurians, many of which only have skeleton consisting of small wheel-like ossicles and jaws reduced to a calcareous ring (Smith and Reich, 2013). Scanning electron micrograph of *Rotasaccus* and *Stichopus chloronotus* courtesy of Mike Reich. Drawing of *Sollasina cthulhu* from Rahman et al. (2019) courtesy of Elissa Sorojsrisom.

leading to the reduction of the holothurian skeleton are well documented in the fossil record (Figure 7) (Smith and Reich, 2013; Rahman et al., 2019). Phylogenetic analyses have consistently identified the ophiocistioids, echinozoans with a mix of characters found in both crown group echinoids and crown group holothurians, as members of the holothurian stem group (Smith, 1988; Smith and Reich, 2013; Rahman et al., 2019). Like many echinoids, most known ophiocistioids in the fossil record have a skeleton comprised of large imbricating $CaCO_3$ plates. In addition to having large embedded plates in their body wall, ophiocistioids also have a jaw apparatus not unlike that of an echinoid's Aristotle's lantern and large plated tube feet, similar to those found in bothriocidaroid echinoids and some somasteroid asterozoans (Jell, 1983; Shackleton, 2005; Reich and Smith, 2009; Thompson et al., 2022). In contrast to echinoids and similar to the crown group holothurians, the Rotasaccidae, a group of ophiocistioids, have reduced much of their skeletons to small wheel-like spicules and have a predominantly soft body wall (Figure 7) (Haude and Langenstrassen, 1976; Reich, 2010). More crown ward stem group holothurians, such as the Devonian *Palaeocucumaria*, also have reduced skeletons and largely unplated bodies, while also having plated tube feet similar to those in the ophiocistioids (Smith and Reich, 2013).

The fossil record provides a clear record of the morphological transitions underlying the evolution of the holothurian body plan (Figure 7). This history also informs on testable hypotheses regarding the genomic and molecular basis for the reduction of the holothurian skeleton. Many of the transcription factors expressed during the development of the echinoderm skeleton seem to be largely conserved across classes and life history stages (Dylus et al., 2018; Erkenbrack and Thompson, 2019). For instance, the transcription factor *Alx1*, known to be a key regulator of downstream biomineralization genes in *S. purpuratus* (Figure 5) (Rafiq et al., 2012), is expressed in the skeletogenic cells of the larval holothurian *Apostichopus parvimensis* (McCauley et al., 2012). Though the larval and adult skeletons are distinct, this suggests that at least some of the transcription factors at the core of the skeletogenic GRN in holothurians are conserved across other eleutherozoan clades. Conversely, comparative analyses of genome content across ambulacrarians have shown that while many of the signaling pathways and transcription factors are conserved, the holothurian *Apostichopus japonicus* has relatively fewer differentiation genes implicated in biomineralization, such as members of the MSP130, C-lectin, and carbonic anhydrase families, than *S. purpuratus*, the asteroid *Acanthaster planci,* or the hemichordate *Saccoglossus kowalevskii* (Zhang et al., 2017). This indicates that the reduction of the skeleton in holothurians, which took place along the holothurian stem lineage in the ophiocistioids (Smith and Reich, 2013; Rahman et al., 2019), may have been underpinned by a reduction in the number of downstream skeletogenic genes, and not associated with the loss or reduced expression of transcription factors.

5.2 What Can We Learn from Comparative Analyses in Extant Taxa?

The key to understanding the evolution of GRNs is to understand conserved and divergent aspects of their topology across differing phylogenetic distances. In order to do this, comparative data on gene expression and gene function is necessary from a wide array of taxa. This is a serious roadblock in evolutionary developmental biology, where the generation of data from within a single taxon takes months and years of, often difficult, experiments. Because the GRN of *S. purpuratus* has been identified in such precise detail, however, comparative studies with other echinoderms were among the first to understand conservation and divergence in GRNs over vast evolutionary distances (Hinman et al., 2003), and echinoderms have become an ideal model system for evolutionary comparisons of GRNs. Gene expression data exists in embryonic or larval development for all five extant classes of echinoderms, and comparisons of gene expression and function across and within these classes are providing a view of GRN evolution at multiple scales within the phylum Echinodermata. As opposed to cross-phylum comparisons, where homologous embryonic structures are difficult to pinpoint, echinoderms fall in a sweet spot. Their embryos are evolutionarily divergent enough to show distinct differences in cell types, gene expression, and morphological structures, yet not too morphologically distinct that cell and tissue types cannot be easily recognized as homologous given multiple criteria. Next I will outline a number of cases where comparative analyses of the gene regulatory basis of echinoderm skeletogenesis have provided insight into the evolution of GRNs in deep time.

5.2.1 Evolution of Divergent Mechanisms of Cell Specification

Comparative analyses of the divergent GRNs across echinoids and other echinoderm outgroups have provided an unparalleled resource with respect to phylogenetic breadth for understanding the pace of and mechanisms underpinning the evolution of development. In euechinoid echinoids, the skeletogenic GRN is activated in the skeletogenic cells through the activity of a GRN subcircuit called the DNG. This molecular mechanism is so-called because it involved the repression of one transcription factor acting as a repressor, by another, resulting in the activation of genes under the control of the second repressor. In particular, in well-studied euechinoids, the expression of key skeletogenic transcription factors is regulated by a double-repression mechanism (Figure 8A) (Revilla-i-Domingo et al., 2007; Oliveri et al., 2008). At the sixteen-cell stage embryo, the transcription factor *Pmar1*, a transcriptional repressor, is expressed in the micromeres, four cells located at the vegetal pole (bottom) of the embryo (Oliveri et al., 2002). Later in development, the transcription factor *HesC*, also a repressor, is expressed in all cells of the embryo, except for those cells where *Pmar1* was expressed earlier in development. Because Pmar1 is a repressor of *HesC*, in those cells where *Pmar1* was expressed earlier in development, *HesC* is not expressed at the later blastula stage. Also during the blastula stage, the key skeletogenic transcription factors *Alx1, Ets1*, and *Tbr,* and the signaling molecule *Delta* are all expressed in the same cells where *Pmar1* was expressed at the sixteen-cell stage (Revilla-i-Domingo et al., 2007;

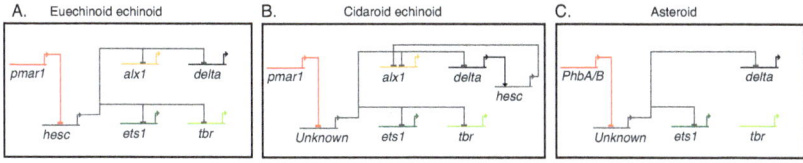

Figure 8 Differences in cell specification mechanisms in the early development
of eleutherozoans.Simplified wiring diagrams showing cell specification
mechanisms in euechinoids, cidaroids, and asteroids. (A) shows the DNG, the
GRN subcircuit through which the skeletogenic cells of numerous euechinoid
echinoids are specified. The repressor Pmar1 represses *HesC*, also a repressor.
HesC represses *Alx1, Ets1, Tbr,* and *Delta,* so the repression of *HesC* by Pmar1
results in the expression of *Alx1, Ets1, Tbr,* and *Delta* later during development
(Revilla-i-Domingo et al., 2007; Oliveri et al., 2008). In cidaroid echinoids (B),
a double-repression mechanism is still present, though *HesC* does not act as
the second repressor. Instead, an unknown gene acts as the second repressor,
repressing *Delta, Ets1,* and *Tbr* (Erkenbrack and Davidson, 2015; Yamazaki
et al., 2020). In asteroids (C), the mechanism even more divergent. This
difference between cidaroids and euechinoids is an example of developmental
systems drift (True and Haag, 2001). Asteroids lack embryonic or larval
skeletogenic cell lineage, though some aspects of this regulatory circuitry are
still present in mesodermal tissue of asteroids. Notably, *Pmar1* is not present in
the genome of asteroids, and its repressive role is fulfilled by PhbA/B
(Yamazaki et al., 2020).

Oliveri et al., 2008). These genes (*Alx1, Ets1, Tbr,* and *Delta*) are under repressive
control of HesC, and thus the activity of Pmar1, which repressed *HesC*, results in
their expression and the specification of the skeletogenic cells that build the
larval skeleton. In contrast to identified euechinoid embryos, aspects of the
mechanism specifying skeletogenic cells in cidaroid echinoids are markedly
different (Yamazaki et al., 2014; Erkenbrack and Davidson, 2015; Yamazaki
et al., 2020). Crucially, HesC does not repress *Ets, Tbr,* or *Delta* (Figure 8B).

The DNG is one of the best characterized GRN subcircuits with respect to
breadth of phylogenetic sampling across the echinoderms and, thus, provides an
ideal model to determine how conserved the genetic regulatory mechanisms
specifying the euechinoid skeletogenic cell actually are, and to determine the
antiquity of this molecular character. Thompson et al. (2017) coded the presence
of the Pmar1–HesC DNG for all echinoderm taxa where the data were available
as of 2017 based on the presence or absence of gene expression, as well as
inferences of regulatory interactions (e.g., a gene acting as a repressor is not

likely to be co-expressed in the same cells as a gene it is repressing). Using time-calibrated phylogenetic trees of echinoids and other echinoderm outgroups, they then used ancestral state reconstruction to infer the probability that the Pmar1–HesC DNG was present or absent at particular ancestral nodes within the echinoids. These analyses showed that the DNG was likely responsible for specifying skeletogenic cells in the MRCA of euechinoid echinoids, and with a lesser probability at the MRCA of crown group echinoids. This work demonstrated that the Pmar1–HesC DNG was probably present in the most recent common ancestor of crown group echinoids; and thus, this particular GRN subcircuit had an origin in at least the late Paleozoic, and has been a largely invariant character throughout the course of crown group echinoid evolution. Recent work by Yamazaki et al. (2020) continued to build on our understanding of the evolution of skeletogenic cell specification. Using transcriptomic analyses, they were able to identify the *Pmar1* gene in the cidaroid *Prionocidaris baculosa* and carried out *over-expression* experiments (injection of excess mRNA into the egg) and knockdown experiments to determine its function. These experiments showed that Pmar1 did not repress *HesC* in cidaroids; however, when *Pmar1* mRNA was injected into *P. baculosa* embryos, *Alx1*, *Ets1,* and *Tbr* all showed upregulation, indicative of a double-repression mechanism. Thus, while there is a double-repression mechanism in cidaroids, the second repressor is not *HesC,* as is the case in *S. purpuratus* and other euechinoids, and its identity remains unknown (Figure 8B). Instead, *HesC* is positively regulated by *Delta,* which is itself under the control of the double-repression mechanism(Yamazaki et al., 2020). In asteroids, where there is no *Pmar1*, or even skeletogenic cell lineage, the regulatory topology present is even more different (Figure 8C) (Cary et al., 2020; Yamazaki et al., 2020).

This case study concerning the evolution of the DNG informs on the mechanisms by which GRNs underlying development evolve more generally. The divergence in developmental mechanisms in cidaroids and euechinoids identified by Yamazaki et al. (2020) is a clear example of the principle of *developmental systems drift* (True and Haag, 2001). Developmental systems drift is the idea that two morphologically homologous structures can develop via divergent molecular or regulatory mechanisms (True and Haag, 2001; Wang and Sommer, 2011). The morphological features that comprise an organism's phenotype are those that interact with the environment; thus, they are under direct selective pressure. The genetic regulatory networks and developmental pathways that encode for morphology, however, are not. As long as the morphological structure remains the same, the molecular pathways expressed during its development, and in particular the regulatory interactions between genes, can vary. This is evident in cidaroids, where an unknown repressor regulates *Delta, Ets1, Tbr,* and *Alx1* in skeletogenic cell specification as opposed to HesC (Figure 8). Because the particular genes that are expressed in development bear little relevance on the morphological outcome, they are able to swap places during the course of evolution. As long as skeletogenic cells are specified, it makes little difference whether HesC or the still unknown repressor is the second repressor of the double-repression mechanism. Within the context of these novel results from Yamazaki et al. (2020), the ancestral state reconstructions of Thompson et al. (2017) also shed light on the timescales over which the effects of developmental systems drift are visible. Comparisons of two nematode species have identified evolutionary changes due to developmental systems drift on timescales of 250–420 Ma (Wang and Sommer, 2011). The taxonomically more expansive analyses of the DNG indicate that in echinoids, the drift from the cidaroid condition of Pmar1-unknown repressor to the euechinoid condition of a Pmar1–HesC double-repression mechanisms likely took place prior to the early Mesozoic, on par with the timescales suggested by nematodes. As comparative analyses in more taxa are carried out, the prevalence of developmental systems drift and the timescales over which its effects are evident will become clearer. Echinoderms, with their wealth of comparative data on gene expression and function, are well suited to play a part.

5.2.2 Skeletogenic Cell Evolution

Tightly tied to the concept of GRNs is the concept of cell types. During development, numerous distinct cell types are specified and differentiate, giving rise to the multitude of tissues, including muscular, nervous, and skeletal, that are present across the body plans of animals. Throughout evolution, novel cell types evolve, giving rise to new tissues, morphological structures, and cellular functions, and leading to both increases and decreases in animal complexity (Valentine et al., 1994; Arendt, 2008; Arendt et al., 2016). While there are numerous definitions for exactly what defines a cell type, the definition used herein is that following Arendt et al. (2016), where a distinct set of transcription factors present in different cells are used to delineate different cell types.

Much controversy has surrounded the origin of the larval skeletons of echinoderms. In particular, the question as to whether or not the larval skeletons of echinoids and ophiuroids are homologous, or the product of convergent evolution, has pervaded the literature (Smith, 1997). The origin of this skeleton can be understood, however, by comparative analyses of the cells that build it. Using a spatial dataset of transcription factor and signaling receptor expression, Erkenbrack and Thompson (2019) used ancestral state reconstructions to infer the likely ancestral states of skeletogenic gene expression in eleutherozoans. These analyses showed that the skeletogenic cells of ophiuroids and echinoids were, in fact, homologous features, based on reconstructed gene expression supporting the expression of *Alx1, Ets1,* and *VegfR* in skeletogenic cells in the eleutherozoan most recent common ancestor (Figure 9). Additionally, the analyses showed that *Tbr*, a transcription factor necessary for skeletogenesis in *S. purpuratus*, only became expressed specifically in skeletogenic cells recently in the evolutionary history of echinoids, in the MRCA of camarodont echinoids (Figure 9). Tbr has far fewer transcriptional inputs into skeletogenic differentiation genes than Ets1 and Alx1 in *S. purpuratus* (Rafiq et al., 2012), and functional knockdown of *Tbr* in non-camarodont echinoids does not appear to effect skeletogenic cell differentiation (Yamazaki et al., 2014). This suggests that the "shallower" regulation of skeletogenic genes by Tbr, as in *S. purpuratus*, may be due to its more evolutionarily recent role in skeletogenesis. This lays out testable predictions for analyses of GRNs in other animal groups, namely: that some more evolutionarily recent additions to GRNs may have transcriptional inputs into fewer downstream genes than more evolutionarily ancient members of those GRNs. This would suggest that evolutionarily older genes in a network might have more time to accumulate new transcriptional targets due to mutations, selection, or drift. This might be expected to happen when functionally redundant or similar differentiation genes in the same network, such as two different genes both involved in skeletogenesis, come under the transcriptional regulation of the same upstream transcription factor due to a mutation in a regulatory element. These insights into the timescale of GRN and cell type evolution can only come about through comparative analyses of multiple taxa spanning wide phylogenetic distances, making echinoderms the ideal model system for evolutionary studies of this kind.

6 Open Questions and Future Directions for Echinoderm Molecular Paleobiology

6.1 Adult Body Plan Development

A current limitation to molecular palaeobiological studies in echinoderms is the disconnect between the vast literature concerning molecular aspects of echinoderm development, and research on the echinoderm fossil record. This disconnect largely exists due to the biphasic lifestyle of echinoderms. While the echinoderm fossil record consists almost entirely of fossilized post-metamorphic or directly developed animals, the majority of studies of developmental gene expression, gene regulation, and protein localization in echinoderms are focused on the embryonic and larval stages of indirect developing echinoderms, which have virtually no fossil record,

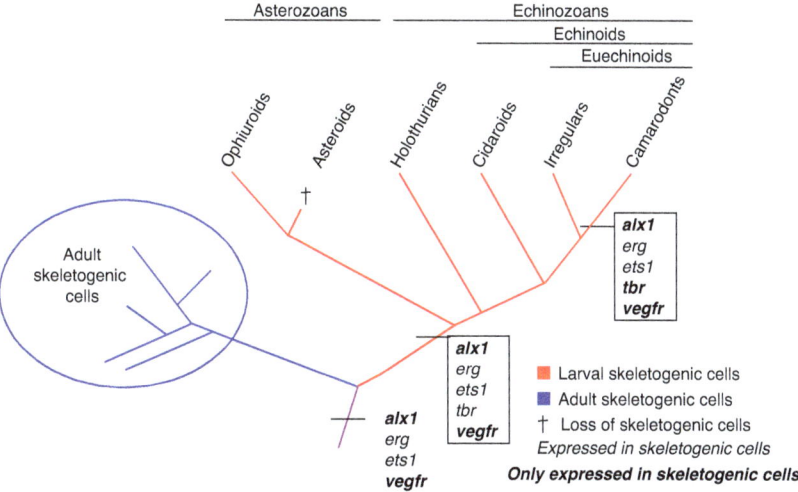

Figure 9 Cell type tree showing hypothesized evolutionary relationships of skeletogenic cells in eleutherozoans. This phylogenetic tree depicts proposed evolutionary relationships among the skeletogenic cells (not taxa) of larval and adult echinoderms. It is presumed that both adult and larval skeletogenic cells of extant echinoderms are descended from the adult skeletogenic cells of extinct echinoderms that expressed skeletogenic genes such as *Alx1, Erg, Ets1,* and the signaling molecule *VegfR*. Through the course of evolution, the transcription factor *Tbr* became co-opted into the skeletogenic process in larval echinoderms, being expressed in skeletogenic cells and other mesodermal tissues. Along the lineage leading to camarodont euechinoids, *Tbr* came to be expressed specifically in the skeletogenic cells, and also became necessary for the normal formation of the larval skeleton. This gradual process leading to ever-more specific expression sheds light onto the processes of gene expression evolution which underlie cell type evolution across the animal kingdom. Figure is modified from Erkenbrack and Thompson (2019).

and a very limited preservation potential. Recent work is attempting to bridge this gap, and work on the development of post-metamorphic and juvenile echinoderms is shedding novel light onto the evolution of the adult body plan.

6.1.1 The Origin of Symmetry

Perhaps the most obvious molecular palaeobiological question to still be answered within the echinoderms is that concerning the developmental and genomic basis of the bizarre, enigmatic, pentaradial symmetry that characterizes members of the echinoderm stem and crown groups. The fossil record tells us that echinoderms have displayed varying forms of symmetry throughout their evolutionary history, from the bilaterally symmetrical *Ctenoimbricata* to the asymmetrical solutes, the triradial helicoplacoids, and the pentaradial forms of more crown ward members (Figure 10A–E) (Zamora and Rahman, 2014). As bilaterians, the ancestral symmetry was likely bilateral, but all extant echinoderms display distinct fivefold symmetry. Though the fossil record displays clear transitions in echinoderm body plans leading to the extant pentaradial forms (Sumrall and Wray, 2007; Zamora and Rahman, 2014), there still remains little understanding of molecular mechanisms underlying the development of the adult body plan in echinoderms.

Crucial to understanding the evolution of echinoderm symmetry is understanding the identity of body axes in echinoderms. During extant echinoderm growth and development, the first morphological structure to show the characteristic fivefold symmetry of echinoderms is the *hydrocoel*, which forms as five lobed outgrowths from one of the coeloms (Morris et al., 2009; Morris, 2011; Morris, 2012). In developing echinoids and asteroids, arrangement of the hydrocoel and other coeloms relative to the mouth supports the idea that the adult oral–aboral (or dorsal–ventral) axis may be equivalent to the anterior–posterior axis of other bilaterians (Peterson et al., 2000; Morris, 2007; Morris et al., 2009; Morris, 2011). During development of the adult body plan, indirect developing echinoderms undergo a coelomic rearrangement, which results in a linear stacking of their coeloms in the adult body plan (Peterson et al., 2000). This stacking results in the location of the left hydrocoel tissues (including the water vascular system) surrounding the mouth, and stacked more adoral than the left and then right *somatocoels*. The expression of posterior *Hox* genes in the somatocoels has been taken as evidence for their posterior identity, which led Peterson et al. (2000) to interpret the anterior–posterior axis as passing from the mouth, through to the stacked coeloms. Thus, this model has the mouth, hydrocoel, left somatocoel, and right somatocoel arranged from anterior to posterior, and interprets the five rays of crown group echinoderms as outgrowths from the anterior–posterior axis analogous to the limbs of arthropods and vertebrates.

Another hypothesis suggests that the metameric organization of echinoderm rays, such as the arms of asterozoans (Czarkwiani et al., 2013) and the ambulacral and interambulacral tissues and plating of echinoids (Morris and Byrne, 2005; Morris, 2009; Morris and Byrne, 2014), are homologous with the single anterior–posterior axis of chordates and other bilaterians (Morris, 2012). This scenario implies that the pentameral arrangement of the echinoderm body plan resulted from the duplication of a single anterior–posterior axis up to five times.

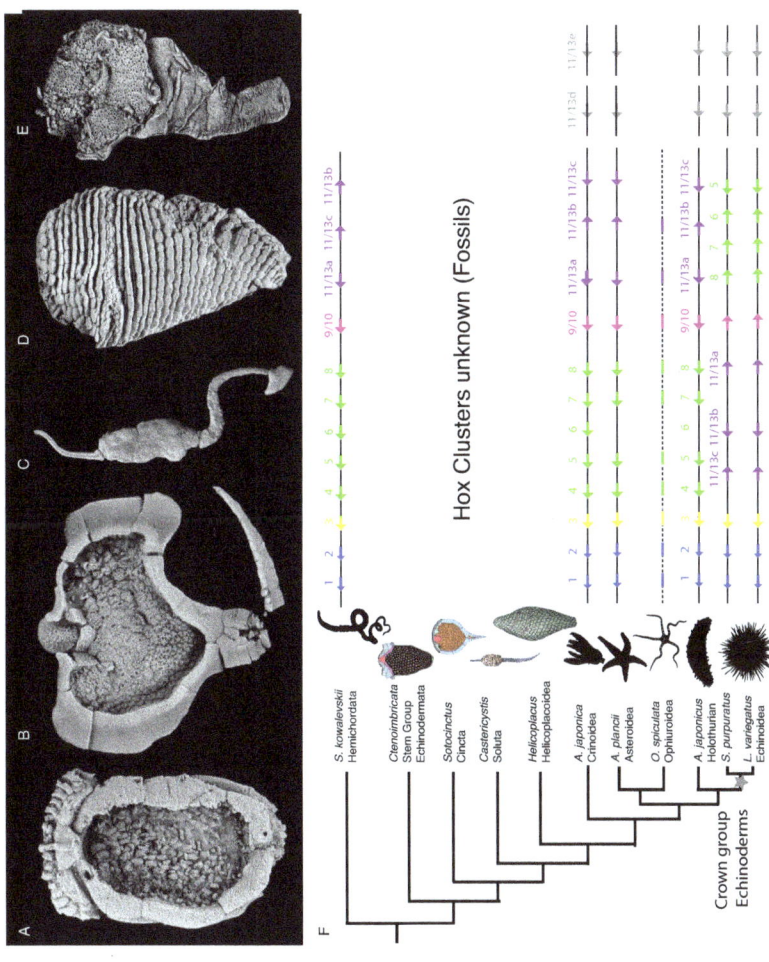

Figure 10 Echinoderm symmetry and the Hox cluster. Various Cambrian echinoderms displaying the varying symmetries present in fossil forms and the organization of the echinoderm Hox cluster. (A) shows the ctenocystoid *Ctenocystis utahensis*. (B) shows the asymmetrical

Caption for Figure 10 (cont.)

cinctan *Lignanicystis barriosensis*. (C) shows the asymmetrical solute *Coleicarpus sprinklei*. (D) shows the triradiate helicoplacoid *Helicoplacus gilberti*. (E) shows the pentaradial *Kinzercystis durhami*. (F) shows the organization of Hox clusters in echinoderms and a hemichordate plotted alongside the hypothesized phylogenetic relationships of ambulacrarians with different symmetries (Zamora and Rahman, 2014). The translocation of the Hox cluster first identified in *S. purpuratus* appears to be an apomorphy unique to echinoids and, thus, cannot be associated with the transition to pentaradial symmetry, which happened earlier in the evolutionary history of echinoids (Byrne et al., 2016). At present this translocation has only been identified in the genomes of regular euechinoids from the order Camarodonta (the grey star on the phylogeny); further work may show it to be present only within this clade or another subclade within Echinoidea. Hox clusters modified from Li et al. (2020) and images and drawings of Palaeozoic echinoderms courtesy of Samuel Zamora. Phylopic images are from Christoph Schomburg (Echinoid), Lauren Sumner-Rooney (Crinoid, Holothurian), Noah Schlottman, photo from Casey Dunn (Ophiuroid; Creative Commons — Attribution-ShareAlike 3.0 Unported — CC BY-SA 3.0), Hans Hillewaert and T. Michael Keesey (Asteroid; Creative Commons — Attribution-ShareAlike 3.0 Unported — CC BY-SA 3.0), and Yan Wong (Hemichordate).

This hypothesis is rooted on the position of a growth zone of terminal addition, from which new axial tissues are added in a metameric fashion during growth (Morris and Byrne, 2005; Morris et al., 2009; Morris and Byrne, 2014). This is similar to the Ocular Plate Rule (OPR), which asserts that new axial tissues grow via terminal addition from a growth zone (Mooi and David, 1994; Mooi et al., 2005), though Mooi et al. (2005) explicitly did not consider manifest-ations of the OPR as examples of metamerism. Building on these interpret-ations, the framework of Minsuk et al. (2009) interpreted echinoderm rays as five proximal–distal axes, as opposed to explicit duplications of the anterior–posterior axis, or outgrowths from a single anterior–posterior axis.

Gene expression patterns and analyses of genomic content and organization have attempted to bring clarity to the question of the echinoderm anterior–posterior axis, and the origin of the pentaradial body plan. Early analyses of the genome of *S. purpuratus* showed that the Hox cluster, the set of closely related transcription factors that are known to specify axial identity along the anterior–posterior axis across a wide array of animal groups (Mallo et al., 2010), had undergone a *translocation* (rearrangement of gene order within along a chromosome) with the order of the Hox genes in the genome rearranged relative to vertebrates, arthropods, and other animals (Martinez et al., 1999; Arenas-Mena et al., 2000; Cameron et al., 2006). Subsequent interpretations have postulated that this translocation of Hox genes relative to their ancestral *collinearity* (matched location of genes on the chromosome relative to their axial expression during development) may have been associated with the pentaradial symmetry in the echinoderm crown group (Mooi and David, 2008; David and Mooi, 2014). One of the first published expression patterns for a hox gene during the formation of the echinoderm adult body plan, *Hox3*, was expressed in a pentaradial pattern in the dental sacs of the echinus rudiment in *S. purpuratus,* and the first high levels of multiple Hox gene expression during *S. purpuratus* development coincide temporally with rudiment forma-tion (Arenas-Mena et al., 1998). The expression of the five posterior-most Hox genes in the somatocoels of *S. purpuratus* revealed a collinear arrangement to their expression in the coelomic mesoderm, where the arrangement of the expression patterns of these genes corresponds with their arrangement in the genome (Arenas-Mena et al., 2000). In *S. purpuratus*, the posterior Hox genes are expressed in the coelomic mesoderm of the left and right somatocoels. The expression patterns of these genes are co-linear, with *Hox11/13b* being expressed in the most posterior tissues of the somatocoel and *Hox7* in the most anterior, in a curved stripe which corresponds to the curvature of the larval gut (Arenas-Mena et al., 2000). The co-linearity of *hox* expression was further identified during development of the crinoid *Metacrinus rotundus*, where

Hox5, *Hox7*, *Hox8*, and *Hox9/10* were found to be expressed in a linear pattern along the length of the somatocoels (Hara et al., 2006). Subsequent work in the direct developing echinoid *Holopneustes purpurescens* showed the expression of the posterior-most Hox gene, *Hox11/13*, in the posterior tissues of the *vestibule*, while the more anterior Hox genes *Hox5* and *Hox3* in more anterior tissues of the epineural folds and coelomic mesoderm, respectively (Morris and Byrne, 2005; Morris and Byrne, 2014). An anterior to posterior expression of Hox genes has also been identified in the somatocoel of the direct developing sand dollar *Peronella japonica* (Tsuchimoto and Yamaguchi, 2014). More recently, the expression patterns of eight Hox genes were surveyed during the pentactula stage of the holothurian *Apostichopus japonicas*, and found to be expressed along the *endodermal* tissues of the digestive tract, albeit in a co-linear fashion as in echinoids and crinoids (Kikuchi et al., 2015).

Though Hox genes have helped to elucidate the orientation of the anterior–posterior axis in echinoderms, their hypothesized role in patterning the pentar-adial body plan of echinoderms has been refuted by genomic data (Figure 10F). As genomic information has become more widespread across echinoderms and advances in sequencing technology have made interrogating echinoderm Hox clusters easier, it has come to light that the translocation of the Hox cluster seen in *S. purpuratus* has not been found in any nonechinoid echinoderms (Baughman et al., 2014; Zhang et al., 2017; Davidson et al., 2020; Li et al., 2020). This may suggest that the translocation of the Hox cluster in echinoids may be a synapomorphy of the class, or at least the camarodont echinoids in which the hox translocation has been identified (Davidson et al., 2020). Given that most echinoderms surveyed have ancestrally ordered hox clusters, the hypothesis that Hox cluster translocation is associated with pentaradiality has confidently been refuted (Figure 10F) (Byrne et al., 2016; Li et al., 2020).

In addition to the Hox genes, the recent work surveying the spatial expression patterns of components of two signaling pathways, the BMP and Nodal pathways, have begun to inform on the development of the pentaradial echinoderm body plan. Nodal is involved in patterning dorsal–ventral axes and left–right asymmetry in numerous animal groups, including larval sea urchins (Molina et al., 2013). In development of the juvenile rudiment of *Heliocidaris erythrogramma*, *Nodal* is expressed in the right ectoderm, while *BMP2/4* was expressed in the left ectoderm in the presumptive vestibular ectoderm, which forms part of the rudiment (Koop et al., 2017). The downstream target of Nodal in embryonic sea urchin development, *BMP2/4*, is also expressed in the hyrdocoel lobes (Koop et al., 2017), which will become sheathed in the vestibule to form the primary podia. Additionally, the transcription factors *Msx*, *Dach*, *Six1/2*, *Six3/6*, and *Pax6*, putative downstream targets of BMP signaling, are expressed in developing hydrocoel lobes, podia, and spines, indicating a role in development of some metameric axial structures of the ambulacral system (Koop et al., 2017; Byrne et al., 2018). Some of these genes are also expressed in the tube feet of the post-metamorphic asteroid *Parvulastra exigua* (Byrne et al., 2020). Though the implication of particular genes in the growth and development of ambulacra is an exceptional step forward in understanding adult and juvenile sea urchin development, these structures are formed after the pentaradial body plan has already been patterned, thus they may not be involved in the process of establishing pentamery (Koop et al., 2017).

6.1.2 The Molecular Basis for Differential Evolvability

Because of their excellent fossil record and molecular resources, echinoderms provide an ideal group to examine differential morphological divergence and constraint in the fossil record, and to attempt to understand the molecular mechanisms underlying differential morphologies in extant taxa. As already mentioned, fossil echinoderms have been a classical model system for understanding morphological disparity, beginning with the work of Foote (1991), Foote (1992) and leading up to more recent treatments from Deline and Ausich (2011), Deline et al. (2020), Hopkins and Smith (2015), and Wright (2017). With the wide array of molecular and genomic tools readily available for echinoderms, including those that can be used to functionally interrogate the development of the adult body, exciting opportunities are developing to understand not only patterns in morphological diversification within the echinoderms, but also the molecular mechanisms that underlie these morphological differences in body plans. An ideal model system for this work is the crown group echinoids. Crown group echinoids provide a prime example of differential morphological divergence and constraint. The regular echinoids, including (among others) cidaroids, camarodonts, and diadematoids, have exhibited striking morphological constraint throughout their evolutionary history, and the earliest regular echinoids in the fossil record appear very morphologically similar to cidaroids in the oceans today (Thompson et al., 2015). In contrast, the irregular echinoids, which evolved from regular echinoid ancestors in the early Jurassic (Saucéde et al., 2007), have undergone extreme morphological diversification since their divergence from the regular echinoids exploring novel morphospace, evolving secondary bilateral symmetry, and exhibiting high morphological disparity (Hopkins and Smith, 2015; Boivin et al., 2018; Mongiardino, 2021; Mongiardino Koch and Thompson, 2021).

Crown group echinoids are the ideal group to examine the molecular underpinning of this differential morphological diversification not only because of the striking differences in morphospace occupation between regular and irregular echinoids, but also because they are the most experimentally tractable model system for functional analyses of adult body plan growth and development. Understanding the differential genomic and cellular mechanisms involved in the development of regular and irregular echinoids will require careful choice and trade-offs between of phylogenetically informative taxa and experimentally tractable animals. Recent experimental progress on the development of the direct developing echinoids with a short time (a matter of days) from fertilization to adult body plan formation such as *H. erythrogramma* (Koop et al., 2017; Edgar et al., 2019a, 2019b; Wang et al., 2020) make these animals prime candidates for understanding the molecular mechanisms underpinning adult body plan growth in regular echinoids. Among the irregular echinoids, the facultatively direct developing clypeasteroid *Clypeaster rosaceus* is a potential ideal choice for experimental work in the post-metamorphic body plan of an irregular echinoid, with readily available transcriptomic resources (Armstrong and Grosberg, 2018) and a relatively short time between fertilization and adult body plan development. Recent work examining skeletogenic cell gene expression in regenerating ophiuroids has shown that different suits of skeletogenic genes are expressed in different skeletal elements of the regenerating arm (Piovani et al., 2021). Whether or not different suites of skeletal genes may also be expressed in differential skeletal elements of regular and irregular echinoids may thus also provide insight into their differential evolvability. With novel hypotheses and experimental organisms, paired with new, robust, functional techniques to examine gene function such as CRISPR Cas-9 genome editing (Liu et al., 2019; Wessel et al., 2020; Yaguchi et al., 2020), uncovering the differential molecular mechanisms underpinning regular and irregular echinoid development and, thus, the vast morphological differences in their morphology are not far off.

7 Concluding Thoughts

Much new data on echinoderm development and evolution has come to light in the sixteen years since Bottjer et al. published "Paleogenomics of Echinoderms." I hope that I have shown herein the utility of echinoderms as an ideal model group for the integration of the fossil record and deep time with data about gene expression, development, and GRNs. Beyond being a tractable model system, novel work from echinoderms are providing fundamental new insight into how cell types, GRNs, and organismal morphology evolve. As comparative approaches involving dense and wide taxonomic sampling, explicit phylogenetic frameworks, and genomic resources become more commonplace in the study of developmental evolution, the rich datasets provided by echinoderms will surely provide even more insight in the coming decades.

Glossary

Animal development – The process by which a fertilized egg grows and undergoes molecular, cellular, and morphological changes throughout the lifetime of the animal.

Archenteron – The invaginated region at the vegetal pole (bottom) of gastrula stage echinoderm embryos. The archenteron forms during gastrulation. The archenteron is the primitive gut, and the opening of the archenteron, the blastopore, will become the anus.

Blastocoel – The open cavity surrounded by cells in a blastula stage embryo.

Blastula stage embryos – A hollow ball of cells that forms following the cleavage stages of embryonic development. In the sea urchin *S. purpuratus*, the blastula stage begins when there are 128 cells in the embryo.

Cell type – A classification used to distinguish different cells from one another in an organism. Cell types can be distinguished based upon morphology, spatial location or position in the anatomy of the species, or based upon molecular characteristics. Recent definitions distinguish different cell types based upon the complement of transcription factors that are expressed within the cell.

Cellular differentiation – The process by which a cell changes from one cell type to another.

Chromatin – DNA and its associated proteins. DNA is stored in the cell as chromatin, which helps to compact it from its total length. DNA is compacted as chromatin in the cell.

Collinearity – Spatial and temporal expression of genes corresponding with the location of these genes on the chromosome and within a genome.

Developmental Gene Regulatory Network – A hierarchical model that represents the numerous regulatory interactions among genes and their products in different spatial and temporal contexts during animal development.

Developmental systems drift – The principle that the genetic networks directing the development of two or more homologous morphological characters can change through the course of evolution, without effecting the morphology of the character. This thus implies a somewhat indirect relationship with genotype and phenotype.

Differentiation genes/proteins – Genes or their resultant proteins that respond to a common set of cell-type specific regulators, including transcription factors, and are responsible for functional and structural characteristics of the cell type. In sea urchins, the SM, or spicule matrix genes, are prime examples of differentiation genes involved in skeletogenesis, as they encode for proteins that are involved in the occluded protein matrix of the skeleton.

Ectoderm – The germ layer located on the outer layer of the embryo. Ectoderm gives rise to tissues of the nervous system, as well as the skin.

Eleutherozoan – The clade consisting of echinoids, holothuroids, ophiuroids, and asteroids.

Endoderm – The endoderm is the innermost germ layer of the embryo. Endodermal derivatives include the epithelium of the gut.

Enhancer – A regulatory module consisting of up to several DNA-binding domains to which transcription factors bind to effect transcription of genes. Enhancers can be located thousands of nucleotides away from the transcription start site and promoter. They are able to effect translation through the activity of DNA looping, whereby the DNA sequences in between the enhancer and the promoter loop, so that the enhancer, and any bound transcription factors, are close to the transcriptional machinery, including the RNA polymerase and other components of the transcription initiation complex.

Epithelial cell – Sheets or tubes of connected cells, originating from any germ layer.

Exon – Exons are the portions of a gene which comprise the final mRNA product during the process of gene expression. They are thus what's left of an mRNA following removal of introns after RNA splicing. It is sometimes said that exons are the portions of the gene which are expressed as they are the portions of the gene which are protein-coding.

Extracellular matrix – Secreted macromolecules immediately surrounding cells. These are secreted by the cells themselves and are useful for cell adhesion and migration.

Filopodial fusion – Fusion of the filopodia, which are long, thin processes extending from the extracellular matrix of cells.

Gastrula stage embryos – Embryos which are undergoing the process of gastrulation, in which multiple layers of the body plan are established. Mesodermal and endodermal cells enter the embryo, while the cells of the ectoderm constitute the outside surface. In indirect developing echinoderm larvae, the archenteron invaginates and the blastopore is formed during the gastrula stage. The gastrula stage follows the blastula stage in indirect developing echinoderm embryos.

Gene – A sequence of nucleotides in DNA or RNA that encodes for a gene product such as an RNA or protein. Gene products have numerous functions throughout animal growth and development, including regulating the expression of other genes, maintaining and carrying out cellular or bio-chemical roles, and synthesizing structures.

Gene duplication – A mutation taking place during the process of DNA replication, which results in the duplication of a segment of DNA along a chromosome. Gene duplication is the process by which paralogues are generated during the course of evolution.

Genome – A genome is the sum of all DNA in an organism. This includes genes that are expressed, but also introns (intragenic regions) and large regions of duplicated and noncoding DNA, which comprise the majority of the genome. It is through differential regulation and expression of portions of the genome that animal development proceeds. Genomes are typically analytically determined through the process of DNA sequencing.

Glycoprotein – Proteins with glycan chains attached via covalent bonds to amino acid side chains. Many extracellular proteins or proteins spanning cell membranes are glycoproteins.

Histone – Small proteins around which DNA is wrapped to make nucleosomes while DNA is compacted in chromatin.

Hydrocoel – One of the coelomic cavaties formed during the development of the adult and post-metamorphic echinoderm body plan. It is from the hydrocoel that the fivefold arrangement of the water vascular system develops.

Induction – An interaction during development by which one set of cells is able to alter the behavior of an adjacent set of cells. This results in changes in cell division rate, morphology, or cell fate. Induction is carried out via cell–cell signalling at close range.

Intron – Introns, short for intragenic regions, are portions of a gene that are noncoding. Introns are removed from mRNA after transcription, a process known as RNA splicing. Introns can be both short and long, and genes can have no introns, or many. In the human genome, most genes are comprised mostly of introns, with the average gene being made up of 95 percent noncoding intragenic regions.

Knockdown – An experimental technique in which the expression of a gene is reduced. By examining the effects on other genes, it is possible to establish evidence for positive or negative gene regulation between the gene that is knocked down, and the other genes. In echinoderms, knockdowns are usually accomplished through the use of morpholino antisense oligo-nucleotides or MASOs which interfere with translation of the protein at the mRNA level.

Ligand – A molecule that binds to a site on a target protein. Ligands are a crucial component of cell–cell signalling pathways. Some are capable of diffusing small distances to bind with receptor proteins during cell–cell communication.

Mesenchyme cell – Loosely packed, unconnected cells, capable of movement and often derived from the mesoderm.

Mesoderm – Mesodermal tissues in echinoderms include muscles, skeleton, gonads, and connective tissues. The mesoderm is located between the ectoderm and endoderm.

Morphogenesis – The process by which cells, tissues, or morphological features of an organism are shaped and arranged into structures during development.

mRNA – Messanger RNA. mRNA is a single-stranded molecule of RNA, which is produced during the process of transcription, during which mRNA is produced from DNA through the action of an RNA polymerase enzyme and other transcriptional machinery. mRNA will typically be translated into a protein, and is thus crucial for monitoring and understanding gene expression during development.

Nucleosome – The coiled structure consisting of DNA, and the histones around which it is wrapped in chromatin. Each nucleosome in eukaryotic cells consists of a core of eight histones, around which DNA is coiled. Nucleosomes are important for the process of gene regulation, as DNA stored as a nucleosomes must be made more (or less) accessible for access by transcription factors.

Over-expression experiments – An experiment when mRNA transcripts corresponding to a particular gene are injected or otherwise introduced into a developing embryo to increase mRNA abundance. Over-expression experiments can be used to understand regulatory interactions. Genes that are upregulated following over-expression of a downstream gene are likely positively regulated by it, while genes whose expression decreases following over-expression of a downstream gene may be under negative regulation by this gene.

Paralog – Homologous genes with shared ancestry resulting from a gene duplication event.

Promoter – A promoter is the site where RNA polymerases and the other transcriptional machinery (including transcription factors) assemble and bind to the DNA during the process of transcription. The promotor is usually upstream of the transcription start site.

Protein – Proteins are macromolecules consisting of chains of amino acids. They are gene products resulting from the translation of mRNA molecules. During development, proteins perform a multiplicity of functions, such as enabling DNA replication and RNA synthesis, regulating the expression of other genes (transcription factors), providing structure, and catalyzing metabolic or other biochemical reactions.

Proteome – The sum of all proteins present in a given organism or tissue.

RNA polymerase – An enzyme capable of synthesizing RNA. In eukaryotes, there are three RNA polymerases, though the one involved transcribing most coding genes is the Polymerase II (Pol II) enzyme. RNA polymerases bind to the promoter of a DNA sequence along with other transcription factors during the process of transcription.

Sclerocytes – Skeletal cells of echinoderms, particularly those not associated with the growth of echinoid teeth.

Signal transduction cascade – Enzymatic reactions taking place within a cell after it receives a signal from another cell. These enzymatic changes can include the phosphorylation of proteins, changes in protein structure, or dimerization (binding of two proteins). The end result of signal transduction cascades is usually the regulation of a transcription factor, which in turn regulates the expression of another gene.

Signaling molecule – Signaling molecules are proteins involved in cell–cell communication. They include ligands and cell-surface proteins that are responsible for sending signals, as well as receptor proteins that receive the signal on another cell. The signal can be sent via direct contact between transmembrane proteins on adjacent cells, or the signal can diffuse over small distances, or be sent via fluids such as blood to travel longer distances. It is through the action of signalling molecules that regulatory action in one cell is capable of inducing changes in gene regulation in other cells.

Somatocoel – Coelomic cavity formed during the development of the adult and post-metamorphic echinoderm body plan. There are two somatocoels: the left somatocoel and the right somatocoel. In indirect developing echinoderms, the adult body plan develops, in part, from tissues of the left somatocoel.

Stereom – The porous, $CaCO_3$ microstructural meshwork which comprises the echinoderm skeleton.

Syncytium – A cell or cytoplasmic mass containing multiple nuclei formed by the fusion of several cells.

Transcription cofactor – A protein that acts in conjunction with another transcription factor, or transcription factors, to regulate the expression of another gene. Transcription cofactors, unlike transcription factors, do not bind directly to DNA, but rather work with other proteins to mediate transcription.

Transcription factor – A transcription factor is a protein, encoded for by a gene, that binds to regulatory sequences on a piece of DNA to regulate the expression of the gene it has bound to. Transcription factors are crucial components of GRNs, as they are responsible for regulating the expression of the genes that they bind to. Transcription factors act combinatorially, and it is the combinatorial activity of transcription factors that results in precise spatial and temporal differences in gene expression during development.

Transcription initiation complex – The set of RNA polymerase and general transcription factors which are bound together at the promotor regulatory sequence of the DNA sequence during transcription of RNA from DNA.

Transcriptional machinery – All of the proteins, enzymes, and other molecules involved in the process of RNA transcription from DNA. This includes any RNA polymerases, general transcription factors, cofactors, co-activators such as the mediator complex, and other proteins which facilitate the process of transcription.

Transcriptome – A transcriptome is the sum of all expressed genes in an organism or tissue. It thus excludes portions of genes which are not expressed (introns) and genes which are not expressed in particular tissues or during certain stages of development. Transcriptomes can thus vary spatially and temporally through the course of animal growth and development. Transcriptomes are typically analytically determined through the process of RNA-sequencing (RNA-seq).

Translocation – A chromosomal mutation in which part of a chromosome breaks and reattaches, resulting in a rearrangement of gene order and location within the genome. The Hox cluster within echinoids has undergone a translocation through the course of their evolutionary history.

Vacuole – Closed membrane-bound sacs within cells filled with water and organic or inorganic molecules in solution.

Vestibule – An ectodermally derived sac that forms over the left hydrocoel of echinoderms prior to metamorphosis. During the development of the adult body plan, the five lobes of the hydrocoel, which will form the ambulacra of the juvenile, elongate through the growing vestibule, and the vestibule forms the ectodermal epithelium of the juvenile's primary podia (tube feet).

References

Adomako-Ankomah, A. & Ettensohn, C. A. 2013. Growth factor-mediated mesodermal cell guidance and skeletogenesis during sea urchin gastrulation. *Development*, 140, 4214–4225.

Adomako-Ankomah, A. & Ettensohn, C. A. 2014. Growth factors and early mesoderm morphogenesis: Insights from the sea urchin embryo. *Genesis*, 52, 158–172.

Ameye, L., Hermann, R., Killian, C., Wilt, F. & Dubois, P. 1999. Ultrastructural localization of proteins involved in sea urchin biomineralization. *Journal of Histochemistry & Cytochemistry*, 47, 1189–1200.

Anstrom, J. A., Chin, J., Leaf, D. S., Parks, A. L. & Raff, R. A. 1987. Localization and expression of msp130, a primary mesenchyme lineage-specific cell surface protein in the sea urchin embryo. *Development*, 101, 255–265.

Arenas-Mena, C., Cameron, A. R. & Davidson, E. H. 2000. Spatial expression of Hox cluster genes in the ontogeny of a sea urchin. *Development*, 127, 4631–4643.

Arenas-Mena, C., Martinez, P., Cameron, R. A. & Davidson, E. H. 1998. Expression of the Hox gene complex in the indirect development of a sea urchin. *Proceedings of the National Academy of Sciences*, 95, 13062–13067.

Arendt, D. 2008. The evolution of cell types in animals: Emerging principles from molecular studies. *Nature Reviews Genetics*, 9, 868–882.

Arendt, D., Musser, J. M., Baker, C. V. et al. 2016. The origin and evolution of cell types. *Nature Reviews Genetics*, 17, 744–757.

Armstrong, A. F. & Grosberg, R. K. 2018. The developmental transcriptomes of two sea biscuit species with differing larval types. *BMC Genomics*, 19, 1–9.

Barsi, J. C., Tu, Q., Calestani, C. & Davidson, E. H. 2015. Genome-wide assessment of differential effector gene use in embryogenesis. *Development*, 142, 3892–3901.

Barsi, J. C., Tu, Q. & Davidson, E. H. 2014. General approach for in vivo recovery of cell type-specific effector gene sets. *Genome Research*, 24, 860–868.

Bauer, J. E. 2021. Paleobiogeography, paleoecology, diversity, and speciation patterns in the Eublastoidea (Blastozoa: Echinodermata). *Paleobiology*, 47, 1–15.

Baughman, K. W., Mcdougall, C., Cummins, S. F. et al. 2014. Genomic organization of Hox and Para Hox clusters in the echinoderm, *Acanthaster planci*. *Genesis*, 52, 952–958.

Boivin, S., Saucède, T., Laffont, R., Steimetz, E. & Neige, P. 2018. Diversification rates indicate an early role of adaptive radiations at the origin of modern echinoid fauna. *Plos One*, 13, e0194575.

Borges, R., Machado, J. P., Gomes, C., Rocha, A. P. & Antunes, A. 2019. Measuring phylogenetic signal between categorical traits and phylogenies. *Bioinformatics*, 35, 1862–1869.

Bottjer, D. J. 2017. Geobiology and palaeogenomics. *Earth Science Reviews*, 164, 182–192.

Bottjer, D. J., Davidson, E. H., Peterson, K. J. & Cameron, R. A. 2006. Paleogenomics of echinoderms. *Science*, 314, 956–960.

Brennand, H. S., Soars, N., Dworjanyn, S. A., Davis, A. R. & Byrne, M. 2010. Impact of ocean warming and ocean acidification on larval development and calcification in the sea urchin *Tripneustes gratilla*. *PloS One*, 5, e11372.

Byrne, M., Ho, M. A., Koleits, L. et al.2013. Vulnerability of the calcifying larval stage of the Antarctic sea urchin *Sterechinus neumayeri* to near-future ocean acidification and warming. *Global Change Biology*, 19, 2264–2275.

Byrne, M., Ho, M., Wong, E. et al. 2011. Unshelled abalone and corrupted urchins: Development of marine calcifiers in a changing ocean. *Proceedings of the Royal Society B: Biological Sciences*, 278, 2376–2383.

Byrne, M., Koop, D., Morris, V. B. et al. 2018. Expression of genes and proteins of the pax-six-eya-dach network in the metamorphic sea urchin: Insights into development of the enigmatic echinoderm body plan and sensory structures. *Developmental Dynamics*, 247, 239–249.

Byrne, M., Koop, D., Strbenac, D. et al. 2020. Transcriptomic analysis of sea star development through metamorphosis to the highly derived pentameral body plan with a focus on neural transcription factors. *DNA Research*, 27, dsaa007.

Byrne, M., Martinez, P. & Morris, V. 2016. Evolution of a pentameral body plan was not linked to translocation of anterior Hox genes: The echinoderm HOX cluster revisited. *Evolution & Development*, 18, 137–143.

Byrne, M. & Selvakumaraswamy, P. 2002. Phylum echinodermata: Ophiuroidea. In Young, C. M., Sewell, M. A. & Rice, M. E. (eds.) *Atlas of marine invertebrate larvae*. San Diego: Academic Press, 483–498.

Cameron, C. & Bishop, C. 2012. Biomineral ultrastructure, elemental constitution and genomic analysis of biomineralization-related proteins in hemichordates. *Proceedings of the Royal Society B: Biological Sciences*, 279, 3041–3048.

Cameron, R. A., Rowen, L., Nesbitt, R. et al. 2006. Unusual gene order and organization of the sea urchin hox cluster. *Journal of Experimental Zoology Part B: Molecular and Developmental Evolution*, 306, 45–58.

Cannon, J. T., Kocot, K. M., Waits, D. S. et al. 2014. Phylogenomic resolution of the hemichordate and echinoderm clade. *Current Biology*, 24, 2827–2832.

Carter, H. F., Thompson, J. R., Elphick, M. R. & Oliveri, P. 2021. The development and neuronal complexity of bipinnaria larvae of the sea star *Asterias rubens*. *Integrative and Comparative Biology*, 61, 337–351.

Cary, G. A., Mccauley, B. S., Zueva, O. et al. 2020. Systematic comparison of sea urchin and sea star developmental gene regulatory networks explains how novelty is incorporated in early development. *Nature communications*, 11(1), 1–9. .

Cheers, M. S. & Ettensohn, C. A. 2005. P16 is an essential regulator of skeletogenesis in the sea urchin embryo. *Developmental Biology*, 283, 384–396.

Chiaramonte, M., Russo, R., Costa, C., Bonaventura, R. & Zito, F. 2020. PI3K inhibition highlights new molecular interactions involved in the skeletogenesis of Paracentrotus lividus embryos. *Biochimica et Biophysica Acta (BBA)-Molecular Cell Research*, 1867, 118558.

Chipman, A. D. & Edgecombe, G. D. 2019. Developing an integrated understanding of the evolution of arthropod segmentation using fossils and evo-devo. *Proceedings of the Royal Society B*, 286, 20191881.

Chow, G. & Benson, S. C. 1979. Carbonic anhydrase activity in developing sea urchin embryos. *Experimental Cell Research*, 124, 451–453.

Clark, E. G., Hutchinson, J. R., Bishop, P. J. & Briggs, D. E. 2020. Arm waving in stylophoran echinoderms: Three-dimensional mobility analysis illuminates cornute locomotion. *Royal Society Open Science*, 7, 200191.

Cole, S. R., Wright, D. F. & Ausich, W. I. 2019. Phylogenetic community paleoecology of one of the earliest complex crinoid faunas (Brechin Lagerstätte, Ordovician). *Palaeogeography, Palaeoclimatology, Palaeoecology*, 521, 82–98.

Costa, C., Karakostis, K., Zito, F. & Matranga, V. 2012. Phylogenetic analysis and expression patterns of p16 and p19 in Paracentrotus lividus embryos. *Development Genes and Evolution*, 222, 245–251.

Croce, J., Lhomond, G., Lozano, J.-C. & Gache, C. 2001. ske-T, a T-box gene expressed in the skeletogenic mesenchyme lineage of the sea urchin embryo. *Mechanisms of Development*, 107, 159–162.

Czarkwiani, A., Dylus, D. V., Carballo, L. & Oliveri, P. 2021. FGF signalling plays similar roles in development and regeneration of the skeleton in the brittle star *Amphiura filiformis*. *Development*, 148(10), dev180760.

Czarkwiani, A., Dylus, D. V. & Oliveri, P. 2013. Expression of skeletogenic genes during arm regeneration in the brittle star *Amphiura filiformis*. *Gene Expression Patterns*, 13, 464–472.

David, B. & Mooi, R. 2014. How Hox genes can shed light on the place of echinoderms among the deuterostomes. *EvoDevo*, 5, 22.

Davidson, E. H. 2006. *The regulatory genome: Gene regulatory networks in development and evolution*, Elsevier.

Davidson, E. H. & Erwin, D. H. 2006. Gene regulatory networks and the evolution of animal body plans. *Science*, 311, 796–800.

Davidson, P. L., Guo, H., Wang, L. et al. 2020. Chromosomal-level genome assembly of the sea urchin *Lytechinus variegatus* substantially improves functional genomic analyses. *Genome Biology and Evolution*, 12, 1080–1086.

Davidson, E. H. & Levine, M. S. 2008. Properties of developmental gene regulatory networks. *Proceedings of the National Academy of Sciences*, 105, 20063–20066.

Davidson, E. H., Rast, J. P., Oliveri, P. et al. 2002a. A genomic regulatory network for development. *Science*, 295, 1669–1678.

Davidson, E. H., Rast, J. P., Oliveri, P. et al. 2002b. A provisional regulatory gene network for specification of endomesoderm in the sea urchin embryo. *Developmental Biology*, 246, 162–190.

Deflandre-Rigaud, M. 1946. Vestiges microscopiques des larves d'echinodermes de l'Oxfordien de Viller-sur-Mer. *CR Acad. Sci. Paris*, 222, 908–910.

Deline, B. & Ausich, W. 2011. Testing the plateau: A reexamination of disparity and morphologic constraints in early Paleozoic crinoids. *Paleobiology*, 37, 214–236.

Deline, B., Thompson, J. R., Smith, N. S. et al. 2020. Evolution and development at the origin of a phylum. *Current Biology*, 30, 1672–1679.

Dubois, P. & Jangoux, M. 1990. Stereom morphogenesis and differentiation during regeneration of adambulacral spines of *Asterias rubens* (Echinodermata, Asteroida). *Zoomorphology*, 109, 263–272.

Duloquin, L., Lhomond, G. & Gache, C. 2007. Localized VEGF signaling from ectodetm to mesenchyme cells controls morphogenesis of the sea urchin embryo skeleton. *Development*, 134, 2293–2302.

Dylus, D. V., Blowes, L. M., Czarkwiani, A., Elphick, M. R. & Oliveri, P. 2018. Developmental transcriptomics of the brittle star *Amphiura filiformis* reveals gene regulatory network rewiring in echinoderm larval skeleton evolution. *Genome Biology*, 19, 26.

Dylus, D. V., Czarkwiani, A., Stångberg, J. et al. 2016. Large-scale gene expression study in the ophiuroid *Amphiura filiformis* provides insights into evolution of gene regulatory networks. *Evodevo*, 7, 1.

Edgar, A., Byrne, M., Mcclay, D. R. & Wray, G. A. 2019a. Evolution of abbreviated development in Heliocidaris erythrogramma dramatically re-wired the highly conserved sea urchin developmental gene regulatory network to decouple signaling center function from ultimate fate. *BioRxiv*, 712216.

Edgar, A., Byrne, M. & Wray, G. A. 2019b. Embryo microinjection of the lecithotrophic sea urchin Heliocidaris erythrogramma. *Journal of Biological Methods*, 6, e119.

Erkenbrack, E. M., Ako-Asare, K., Miller, E. et al. 2016. Ancestral state reconstruction by comparative analysis of a GRN kernel operating in echinoderms. *Development Genes and Evolution*, 226, 37–45.

Erkenbrack, E. M. & Davidson, E. H. 2015. Evolutionary rewiring of gene regulatory network linkages at divergence of the echinoid subclasses. *Proceedings of the National Academy of Sciences*, 112, E4075–E4084.

Erkenbrack, E. M. & Petsios, E. 2017. A conserved role for VEGF signaling in specification of homologous mesenchymal cell types positioned at spatially distinct developmental addresses in early development of sea urchins. *Journal of Experimental Zoology Part B: Molecular and Developmental Evolution*, 328, 423–432.

Erkenbrack, E. M. & Thompson, J. R. 2019. Cell type phylogenetics informs the evolutionary origin of echinoderm larval skeletogenic cell identity. *Communications Biology*, 2, 160.

Erwin, D. H. 2020. The origin of animal body plans: A view from fossil evidence and the regulatory genome. *Development*, 147, dev182899.

Erwin, D. H. & Davidson, E. H. 2009. The evolution of hierarchical gene regulatory networks. *Nature Reviews Genetics*, 10, 141–148.

Ettensohn, C. A. 2009. Lessons from a gene regulatory network: Echinoderm skeletogenesis provides insights into evolution, plasticity and morphogenesis. *Development*, 136, 11–21.

Ettensohn, C. A. 2014. Horizontal transfer of the msp130 gene supported the evolution of metazoan biomineralization. *Evolution & Development*, 16, 139–148.

Ettensohn, C. A. & Dey, D. 2017. KirrelL, a member of the lg-domain superfamily of adhesion proteins, is essential for fusion of primary mesenchyme cells in the sea urchin embryo. *Developmental Biology*, 421, 258–270.

Ettensohn, C. A., Illies, M. R., Oliveri, P. & De jong, D. L. 2003. Alx1, a member of the Cart1/Alx3/Alx4 subfamily of Paired-class homeodomain proterins, is an essential component of the gene regulatory network controlling skeletogenic fate specification in the sea urchin embryo. *Development*, 130, 2917–2928.

Fleming, J. F., Kristensen, R. M., Sørensen, M. V. et al. 2018. Molecular palaeontology illuminates the evolution of ecdysozoan vision. *Proceedings of the Royal Society B*, 285, 20182180.

Foote, M. 1991. Morphological and Taxonomic diversity in a clade's history: The blastoid record and stochastic simulations. *Contributions from the museum of Paleontology The University of Michigan*, 28, 101–140.

Foote, M. 1992. Paleozoic record of morphological diversity in blastozoan echinoderms. *Proceedings of the National Academy of Sciences of the United States of America*, 89, 7325–7329.

Gao, F. & Davidson, E. H. 2008. Transfer of a large gene regulatory apparatus to a new developmental address in echinoid evolution. *Proceedings of the National Academy of Sciences*, 105, 6091–6096.

Gao, F., Thompson, J. R., Petsios, E. et al. 2015. Juvenile skeletogenesis in anciently diverged sea urchin clades. *Developmental Biology*, 400, 148–158.

Garner, S., Zysk, I., Byrne, G. et al. 2016. Neurogenesis in sea urchin embryos and the diversity of deuterostome neurogenic mechanisms. *Development*, 143, 286–297.

Garwood, R. J., Sharma, P. P., Dunlop, J. A. & Giribet, G. 2014. A Paleozoic stem group to mite harvestmen revealed through integration of phylogenetics and development. *Current Biology*, 24, 1017–1023.

Gilbert, S. F. 2006. *Developmental biology*. 8th ed. Sunderland, MA: Sinauer Associates.

Gliznutsa, L. & Dautov, S. S. 2011. Cell differentiation during the larval development of the ophiuroid *Amphipholis kochii* Lütken, 1872 (Echinodermata: Ophiuroidea). *Russian Journal of Marine Biology*, 37, 384–400.

Gorzelak, P. 2021. Functional micromorphology of the echinoderm skeleton. *Cambridge Elements*.

Grun, T. B. & Nebelsick, J. H. 2018. Structural design of the minute clypeasteroid echinoid *Echinocyamus pusillus*. *Royal Society Open Science*, 5, 171323.

Guss, K. & Ettensohn, C. A. 1997a. Skeletal morphogenesis in the sea urchin embryo: Regulation of primary mesenchyme gene expression and skeletal rod growth by ectoderm-derived cues. *Development*, 124, 1899–1908.

Guss, K. A. & Ettensohn, C. A. 1997b. Skeletal morphogenesis in the sea urchin embryo: Regulation of primary mesenchyme gene expression and skeletal rod growth by ectoderm-derived cues. *Development*, 124, 1899–1908.

Hara, Y., Yamaguchi, M., Akasaka, K. et al. 2006. Expression patterns of Hox genes in larvae of the sea lily *Metacrinus rotundus*. *Development Genes and Evolution*, 216, 797–809.

Hart, M. W., Byrne, M. & Smith, M. J. 1997. Molecular phylogenetic analysis of life-history evolution in asterinid starfish. *Evolution*, 51, 1848–1861.

Haude, R. & Langenstrassen, F. 1976. *Rotasaccus dentifer* n. g. n. sp., ein devonischer Ophiocistioide (Echinodermata) mit "holothuroiden" Wandskleriten und "echinoidem" Kauapparat. *Paläontologische Zeitschrift*, 50, 130–150.

Heatfield, B. M. & Travis, D. F. 1975. Ultrastructural studies of regenerating spines of the sea urchin *Strongylocentrotus purpuratus* I. Cell types without spherules. *Journal of Morphology*, 145, 13–49.

Hinman, V. F., Nguyen, A. T., Cameron, R. A. & Davidson, E. H. 2003. Developmental gene regulatory network architecture across 500 million years of echinoderm evolution. *Proceedings of the National Academy of Sciences of the United States of America*, 100, 13356–13361.

Hopkins, M. J. & Smith, A. B. 2015. Dynamic evolutionary change in post-Paleozoic echinoids and the importance of scale when interpreting changes in rates of evolution. *Proceedings of the National Academy of Sciences*, 112, 3758–3763.

Howard, R. J., Puttick, M. N., Edgecombe, G. D. & Lozano-Fernandez, J. 2020. Arachnid monophyly: Morphological, palaeontological and molecular support for a single terrestrialization within Chelicerata. *Arthropod Structure & Development*, 59, 100997.

Jell, P. A. 1983. Early Devonian echinoderms from Victoria (Rhombifera, Blastoidea and Ophiocistioidea). *Memoir of the Association of Australasian Palaeontologists*, 1, 209–235.

Khor, J. M. & Ettensohn, C. A. 2017. Functional divergence of paralogous transcription factors supported the evolution of biomineralization in echinoderms. *Elife*, 6, e32728.

Khor, J. M. & Ettensohn, C. A. 2020. Transcription factors of the Alx family: Evolutionarily conserved regulators of deuterostome skeletogenesis. *Frontiers in Genetics*, 11, 1405.

Khor, J. M., Guerrero-Santoro, J. & Ettensohn, C. A. 2019. Genome-wide identification of binding sites and gene targets of Alx1, a pivotal regulator of echinoderm skeletogenesis. *Development*, 146, dev180653.

Kikuchi, M., Omori, A., Kurokawa, D. & Akasaka, K. 2015. Patterning of antero-posterior body axis displayed in the expression of Hox genes in sea cucumber *Apostichopus japonicus*. *Development Genes and Evolution*, 225, 275–286.

Killian, C. E. & Wilt, F. H. 2008. Molecular aspects of biomineralization of the echinoderm endoskeleton. *Chemical Reviews*, 108, 4463–4474.

Koga, H., Fujitani, H., Morino, Y. et al. 2016. Experimental approach reveals the role of alx1 in the evolution of the echinoderm larval skeleton. *PLoS One*, 11, e0149067.

Koga, H., Matsubara, M., Fujitani, H. et al. 2010. Functional evolution of Ets in echinoderms with focus on the evolution of echinoderm larval skeletons. *Development Genes and Evolution*, 220, 107–115.

Koga, H., Morino, Y. & Wada, H. 2014. The echinoderm larval skeleton as a possible model system for experimental evolutionary biology. *Genesis*, 52, 186–192.

Koop, D., Cisternas, P., Morris, V. B. et al. 2017. Nodal and BMP expression during the transition to pentamery in the sea urchin *Heliocidaris erythrogramma*: Insights into patterning the enigmatic echinoderm body plan. *BMC Developmental Biology*, 17, 1–13.

Kurokawa, D., Kitajima, T., Mitsunaga-Nakatsubo, K. et al. 1999. HpEts, an ets-related transcription factor implicated in primary mesenchyme cell differentiation in the sea urchin embryo. *Mechanisms of Development*, 80, 41–52.

Leaf, D. S., Anstrom, J. A., Chin, J. E. et al. 1987. Antibodies to a fusion protein identify a cDNA clone encoding msp130, a primary mesenchyme-specific cell surface protein of the sea urchin embryo. *Developmental Biology*, 121, 29–40.

Levine, M. & Davidson, E. H. 2005. Gene regulatory networks for development. *Proceedings of the National Academy of Sciences*, 102, 4936–4942.

Li, Y., Omori, A., Flores, R. L. et al. 2020. Genomic insights of body plan transitions from bilateral to pentameral symmetry in Echinoderms. *Communications Biology*, 3, 1–10.

Liu, D., Awazu, A., Sakuma, T., Yamamoto, T. & Sakamoto, N. 2019. Establishment of knockout adult sea urchins by using a CRISPR-Cas9 system. *Development, Growth & Differentiation*, 61, 378–388.

Livingston, B. T., Killian, C. E., Wilt, F. H. et al. 2006. A genome-wide analysis of biomineralization-related proteins in the sea urchin *Strongylocentrotus purpuratus*. *Developmental Biology*, 300, 335–348.

Lozano-Fernandez, J., Carton, R., Tanner, A. R. et al. 2016. A molecular palaeobiological exploration of arthropod terrestrialization. *Philosophical Transactions of the Royal Society B: Biological Sciences*, 371, 20150133.

Mallo, M., Wellik, D. M. & DESCHAMPS, J. 2010. Hox genes and regional patterning of the vertebrate body plan. *Developmental Biology*, 344, 7–15.

Mann, K., Poustka, A. J. & Mann, M. 2008a. In-depth, high-accuracy proteomics of sea urchin tooth organic matrix. *Proteome Science*, 6, 33.

Mann, K., Poustka, A. J. & Mann, M. 2008b. The sea urchin (*Strongylocentrotus purpuratus*) test and spine proteomes. *Proteome Science*, 6, 1–10.

Mann, K., Wilt, F. H. & Poustka, A. J. 2010. Proteomic analysis of sea urchin (*Strongylocentrotus purpuratus*) spicule matrix. *Proteome Science*, 8, 1–12.

Marie, B., Zanella-Cléon, I., Guichard, N., Becchi, M. & Marin, F. 2011. Novel proteins from the calcifying shell matrix of the Pacific oyster Crassostrea gigas. *Marine Biotechnology*, 13, 1159–1168.

Märkel, K., Röser, U., Mackenstedt, U. & Klostermann, M. 1986. Ultrastructural investigation of matrix-mediated biomineralization in echinoids (Echinodermata, Echinoida). *Zoomorphology*, 106, 232–243.

Märkel, K., Röser, U. & Stauber, M. 1989. On the ultrastructure and the supposed function of the mineralizing matrix coat of sea urchins (Echinodermata, Echinoida). *Zoomorphology*, 109, 79–87.

Martinez, P., Rast, J. P., Arenas-Mena, C. & Davidson, E. H. 1999. Organization of an echinoderm Hox gene cluster. *Proceedings of the National Academy of Sciences*, 96, 1469–1474.

Mccauley, B. S., Wright, E. P., Exner, C., Kitazawa, C. & Hinman, V. F. 2012. Development of an embryonic skeletogenic mesenchyme lineage in a sea cucumber reveals the trajectory of change for the evolution of novel structures in echinoderms. *EvoDevo*, 3, 17.

Mcclay, D. R., Warner, J., Martik, M., Miranda, E. & Slota, L. 2020. Gastrulation in the sea urchin. *Current Topics in Developmental Biology*, 136, 195–218.

Mcedward, L. R. & Miner, B. G. 2001. Larval and life-cycle patterns in echinoderms. *Canadian Journal of Zoology*, 79, 1125–1170.

Minokawa, T., Hamaguchi, Y. & Amemiya, S. 1997. Skeletogenic potential of induced secondary mesenchyme cells derived from the presumptive ectoderm in echinoid embryos. *Development Genes and Evolution*, 206, 472–476.

Minsuk, S. B., Turner, F. R., Andrews, M. E. & Raff, R. A. 2009. Axial patterning of the pentaradial adult echinoderm body plan. *Development Genes and Evolution*, 219, 89–101.

Mitsunaga, K., Akasaka, K., Shimada, H. et al. 1986. Carbonic anhydrase activity in developing sea urchin embryos with special reference to calcification of spicules. *Cell Differentiation*, 18, 257–262.

Molina, M. D., De Crozé, N., Haillot, E. & Lepage, T. 2013. Nodal: Master and commander of the dorsal–ventral and left–right axes in the sea urchin embryo. *Current Opinion in Genetics & Development*, 23, 445–453.

Mongiardino Koch, N., Coppard, S. E., Lessios, H. A. et al. 2018. A phylogenomic resolution of the sea urchin tree of life. *BMC Evolutionary Biology*, 18, 189.

Mongiardino Koch, N. (2021). Exploring adaptive landscapes across deep time: A case study using echinoid body size. *Evolution*, 75(6), 1567–1581.

Mongiardino Koch, N. & Thompson, J. R. 2021. A Total-evidence dated phylogeny of echinoidea combining phylogenomic and paleontological data. *Systematic Biology*, 70, 421–439

Mooi, R. & David, B. 1994. Echinoderm skeletal homologies: Classical morphology meets modern phylogenetics. In David, B., Guille, A., Féral, J.-P. & Roux, M. (eds.) *Echinoderms through time*. Rotterdam: A. A. Balkema, 87–95.

Mooi, R. & David, B. 2008. Radial symmetry, the anterior/posterior axis, and echinoderm Hox genes. *Annual Review of Ecology, Evolution, and Systematics*, 39, 43–62.

Mooi, R., David, B. & Wray, G. A. 2005. Arrays in rays: Terminal addition in echinoderms and its correlation with gene expression. *Evolution & Development*, 7, 542–555.

Morgulis, M., Gildor, T., Roopin, M. et al. 2019. Possible cooption of a VEGF-driven tubulogenesis program for biomineralization in echinoderms. *Proceedings of the National Academy of Sciences*, 116, 12353–12362.

Morino, Y., Koga, H., Tachibana, K. et al. 2012. Heterochronic activation of VEGF signaling and the evolution of the skeleton in echinoderm pluteus larvae. *Evolution & Development*, 14, 428–436.

Morris, V. B. 2007. Origins of radial symmetry identified in an echinoderm during adult development and the inferred axes of ancestral bilateral symmetry. *Proceedings of the Royal Society B: Biological Sciences*, 274, 1511–1516.

Morris, V. B. 2009. On the sites of secondary podia formation in a juvenile echinoid: Growth of the body types in echinoderms. *Development Genes and Evolution*, 219, 597–608.

Morris, V. B. 2011. Coelomogenesis during the abbreviated development of the echinoid *Heliocidaris erythrogramma* and the developmental origin of the echinoderm pentameral body plan. *Evolution & Development*, 13, 370–381.

Morris, V. B. 2012. Early development of coelomic structures in an echinoderm larva and a similarity with coelomic structures in a chordate embryo. *Development Genes and Evolution*, 222, 313–323.

Morris, V. B. & Byrne, M. 2005. Involvement of two Hox genes and Otx in echinoderm body-plan morphogenesis in the sea urchin *Holopneustes purpurescens*. *Journal of Experimental Zoology Part B: Molecular and Developmental Evolution*, 304, 456–467.

Morris, V. B. & Byrne, M. 2014. Oral–aboral identity displayed in the expression of HpHox3 and HpHox11/13 in the adult rudiment of the sea urchin *Holopneustes purpurescens*. *Development Genes and Evolution*, 224, 1–11.

Morris, V. B., Selvakumaraswamy, P., Whan, R. & Byrne, M. 2009. Development of the five primary podia from the coeloms of a sea star larva: Homology with the echinoid echinoderms and other deuterostomes. *Proceedings of the Royal Society B: Biological Sciences*, 276, 1277–1284.

Mortensen, T. 1921. *Studies of the development and larval forms of echinoderms*, Copenhagen: G. E. C. GAD.

O'hara, T. D., Hugall, A. F., Thuy, B. & Moussalli, A. 2014. Phylogenomic resolution of the class Ophiuroidea unlocks a global microfossil record. *Current Biology*, 24, 1874–1879.

Okazaki, K. 1975. Spicule formation by isolated micromeres of the sea urchin embryo. *American Zoologist*, 15, 567–581.

Oliveri, P., Carrick, D. M. & Davidson, E. H. 2002. A regulatory gene network that directs micromere specification in the sea urchin embryo. *Developmental Biology*, 246, 209–228.

Oliveri, P., Tu, Q. & Davidson, E. H. 2008. Global regulatory logic for specification of an embryonic cell lineage. *Proceedings of the National Academy of Sciences*, 105, 5955–5962.

Pagel, M. 1999. Inferring the historical patterns of biological evolution. *Nature*, 401, 877–884.

Pennington, J. T. & Strathmann, R. R. 1990. Consequences of the calcite skeletons of planktonic echinoderm larvae for orientation, swimming, and shape. *The Biological Bulletin*, 179, 121–133.

Peter, I. S. & Davidson, E. H. 2011. Evolution of gene regulatory networks controlling body plan development. *Cell*, 144, 970–985.

Peter, I. S. & Davidson, E. H. 2015. *Genomic control process: Development and evolution*, London: Academic Press.

Peter, I. S. & Davidson, E. H. 2016. Implications of developmental gene regulatory networks inside and outside developmental biology. *Current Topics in Developmental Biology*, 117, 237–251.

Peter, I. S. & Davidson, E. H. 2017. Assessing regulatory information in developmental gene regulatory networks. *Proceedings of the National Academy of Sciences*, 114, 5862–5869.

Peterson, K. J., Arenas-Mena, C. & Davidson, E. H. 2000. The A/P axis in echinoderm ontogeny and evolution: Evidence from fossils and molecules. *Evolution & Development*, 2, 93–101.

Peterson, K. J., Summons, R. E. & Donoghue, P. C. J. 2007. Molecular palaeobiology. *Palaeontology*, 50, 775–809.

Piovani, L., Czarkwiani, A., Ferrario, C., Sugni, M. & Oliveri, P. 2021. Ultrastructural and molecular analysis of the origin and differentiation of cells mediating brittle star skeletal regeneration. *BMC Biology*, 19, 1–19.

Raff, R. & Byrne, M. 2006. The active evolutionary lives of echinoderm larvae. *Heredity*, 97, 244–252.

Rafiq, K., Cheers, M. S. & Ettensohn, C. A. 2012. The genomic regulatory control of skeletal morphogenesis in the sea urchin. *Development*, 139, 579–590.

Rafiq, K., Shashikant, T., Mcmanus, C. J. & Ettensohn, C. A. 2014. Genome-wide analysis of the skeletogenic gene regulatory network of sea urchins. *Development*, 141, 950–961.

Rahman, I. A., Thompson, J. R., Briggs, D. E. et al. 2019. A new ophiocistioid with soft-tissue preservation from the Silurian Herefordshire Lagerstätte, and the evolution of the holothurian body plan. *Proceedings of the Royal Society B*, 286, 20182792.

Reich, M. 2010. Evolution and diversification of ophiocistioids (Echinodermata: Echinozoa). In Harris, L. G., Böttger, S. A., Walker, C. W. & Lesser, M. P. (eds.) *Echinoderms: Durham*. London: Taylor & Francis, 51–54.

Reich, M. 2021. The first Cretaceous ophiopluteus skeleton (Echinodermata: Ophiuroidea). *Journal of Paleontology*, 95(6), 1284–1292.

Reich, M. & Smith, A. B. 2009. Origins and biomechanical evolution of teeth in echinoids and their relatives. *Palaeontology*, 52, 1149–1168.

Revilla-I-Domingo, R., Oliveri, P. & Davidson, E. H. 2007. A missing link in the sea urchin embryo gene regulatory network: *HesC* and hte double-negative specification of micromeres. *Proceedings of the National Academy of Sciences*, 104, 12383–12388.

Riedl, R. 1977. A systems-analytical approach to macro-evolutionary phenomena. *The Quarterly Review of Biology*, 52, 351–370.

Rizzo, F., Fernandez-Serra, M., Squarzoni, P., Archimandritis, A. & Arnone, M. I. 2006. Identification and developmental expression of the *ets* gene family in the sea urchin (*Strongylocentrotus purpuratus*). *Developmental Biology*, 300, 35–48.

Röttinger, E., Saudemont, A., Duboc, V. et al. 2008. FGF signals guide migration of mesenchymal cells, controls skeletal morphogenesis and regulate gastrulation during sea urchin development. *Development*, 137, 353–365.

Russo, R., Pinsino, A., Costa, C. et al. 2014. The newly characterized Pl-jun is specifically expressed in skeletogenic cells of the *Paracentrotus lividus* sea urchin embryo. *The FEBS Journal*, 281, 3828–3843.

Saucéde, T., Mooi, R. & David, B. 2007. Phylogeny and origin of Jurassic irregular echinoids (Echinodermata: Echinoidea). *Geological Magazine*, 144, 333–359.

Schirrmeister, B. E., Gugger, M. & Donoghue, P. C. 2015. Cyanobacteria and the great oxidation event: Evidence from genes and fossils. *Palaeontology*, 58, 769–785.

Schoch, R. R. 2010. Riedl's burden and the body plan: Selection, constraint, and deep time. *Journal of Experimental Zoology Part B: Molecular and Developmental Evolution*, 314, 1–10.

Sewell, M. A. & Mceuen, F. S. 2002. Phylum echinodermata: Holothuroidea. In Young, C. M., Sewell, M. A. & Rice, M. E. (eds.) *Atlas of marine invertebrate larvae*. San Diego: Academic Press, 513–530.

Shackleton, J. D. 2005. Skeletal homologies, phylogeny and classification of the earliest asterozoan echinoderms. *Journal of Systematic Palaeontology*, 3, 29–114.

Sharma, T. & Ettensohn, C. A. 2010. Activation of the skeletogenic gene regulatory network in the early sea urchin embryo. *Development*, 137, 1149–1157.

Shashikant, T., Khor, J. M. & Ettensohn, C. A. 2018. From genome to anatomy: The architecture and evolution of the skeletogenic gene regulatory network of sea urchins and other echinoderms. *Genesis*, 56, e23253.

Shimizu, M. 1997. Cellular elements of the test plates in the sea urchin *stronglyocentrotus intermedius*. *Fisheries Science*, 63, 161–168.

Shubin, N., Tabin, C. & Carroll, S. 1997. Fossils, genes and the evolution of animal limbs. *Nature*, 388, 639–648.

Shubin, N., Tabin, C. & Carroll, S. 2009. Deep homology and the origins of evolutionary novelty. *Nature*, 457, 818–823.

Smith, A. B. 1980. Stereom microstructure of the echinoid test. *Special Papers in Palaeontology*, 25, 1–85.

Smith, A. B. 1988. Fossil evidence for the relationships of extant echinoderm classes and their times of divergence. In Paul, C. R. C. & Smith, A. B. (eds.) *Echinoderm phylogeny and evolutionary biology*. Oxford: Clarendon Press, 85–97.

Smith, A. B. 1990. Biomineralization in echinoderms. In Carter, J. G. (ed.) *Skeletal biomineralization: Patterns, processes and evolutionary trends. Volume 1*. New York: Van Nostrand Reinhold, 413–442.

Smith, A. B. 1997. Echinoderm larvae and phylogeny. *Annual Review of Ecology and Systematics*, 28, 219–241.

Smith, A. B. & Reich, M. 2013. Tracing the evolution of the holothurian body plan through stem-group fossils. *Biological Journal of the Linnean Society*, 109, 670–681.

Sodergren, E., Weinstock, G. M., Davidson, E. H. et al. 2006. The genome of the sea urchin Strongylocentrotus purpuratus. *Science*, 314, 941–952.

Sperling, E. A., Pisani, D. & Peterson, K. J. 2011. Molecular paleobiological insights into the origin of the Brachiopoda. *Evolution & Development*, 13, 290–303.

Strathmann, R. & Eernisse, D. 1994. What molecular phylogenies tell us about the evolution of larval forms. *American Zoologist*, 34, 502–512.

Strathmann, R. R. 1971. The feeding behavior of planktotrophic echinoderm larvae: Mechanisms, regulation, and rates of suspensionfeeding. *Journal of Experimental Marine Biology and Ecology*, 6, 109–160.

Strathmann, R. R. 1975. Larval feeding in echinoderms. *American Zoologist*, 15, 717–730.

Stricker, S. A. 1985. The ultrastructure and formation of the calcareous ossicles in the body wall of the sea cucumber Leptosynapta clarki (Echinodermata, Holothuroida). *Zoomorphology*, 105, 209–222.

Stricker, S. A. 1986. The fine structure and development of calcified skeletal elements in the body wall of holothurian echinoderms. *Journal of Morphology*, 188, 273–288.

Sumrall, C. D. & Wray, G. A. 2007. Ontogeny in the fossil record: Diversification of body plans and the evolution of "aberrant" symmetry in Paleozoic echinoderms. *Paleobiology*, 33, 149–163.

Syverson, V. J. & Baumiller, T. K. 2014. Temporal trends of predation resistance in Paleozoic crinoid arm branching morphologies. *Paleobiology*, 40, 417–427.

Szabó, R. & Ferrier, D. E. 2015. Another biomineralising protostome with an msp130 gene and conservation of msp130 gene structure across Bilateria. *Evolution & Development*, 17, 195–197.

Telford, M. J., Lowe, C. J., Cameron, C. B. et al. 2014. Phylogenomic analysis of echinoderm class relationships supports Asterozoa. *Proceedings of the Royal Society B: Biological Sciences*, 281, 20140479.

Thompson, J. R., Erkenbrack, E. M., Hinman, V. F. et al. 2017. Paleogenomics of echinoids reveals an ancient origin for the double-negative specification of micromeres in sea urchins. *Proceedings of the National Academy of Sciences*, 114, 5870–5877.

Thompson, J. R., Mirantsev, G. V., Petsios, E. & Bottjer, D. J. 2020. Phylogenetic analysis of the archaeocidaridae and palaeozoic miocidaridae (Echinodermata: Echinoidea) and the origin of crown group echinoids. *Papers in Palaeontology*, 6, 217–249.

Thompson, J. R., Paganos, P., Benvenuto, G., Arnone, M. I. & Oliveri, P. 2021. Post-metamorphic skeletal growth in the sea urchin *Paracentrotus lividus* and implications for body plan evolution. *EvoDevo*, 12, 1–14.

Thompson, J. R., Petsios, E., Davidson, E. H. et al. 2015. Reorganization of sea urchin gene regulatory networks at least 268 million years ago as revealed by oldest fossil cidaroid echinoid. *Scientific Reports*, 5, 1–9.

Thompson, J. R., Cotton, L. J., Candela, Y., Kutscher, M., Reich, M., & Bottjer, D. J. 2022. The Ordovician diversification of sea urchins: systematics of the Bothriocidaroida (Echinodermata: Echinoidea). *Journal of Systematic Palaeontology*, 19(20), 1395–1448.

True, J. R. & Haag, E. S. 2001. Developmental system drift and flexibility in evolutionary trajectories. *Evolution & Development*, 3, 109–119.

Tsuchimoto, J. & Yamaguchi, M. 2014. Hox expression in the direct-type developing sand dollar *Peronella japonica*. *Developmental Dynamics*, 243, 1020–1029.

Tweedt, S. M. 2017. Gene regulatory networks, homology, and the early panarthropod fossil record. *Integrative and Comparative Biology*, 57, 477–487.

Valentine, J. W. & Campbell, C. A. 1975. Genetic regulation and the fossil record: Evolution of the regulatory genome may underlie the rapid development of major animal groups. *American Scientist*, 63, 673–680.

Valentine, J. W., Collins, A. G. & Meyer, C. P. 1994. Morphological complexity increase in metazoans. *Paleobiology*, 20, 131–142.

Vinther, J., Sperling, E. A., Briggs, D. E. & Peterson, K. J. 2012. A molecular palaeobiological hypothesis for the origin of aplacophoran molluscs and their derivation from chiton-like ancestors. *Proceedings of the Royal Society B: Biological Sciences*, 279, 1259–1268.

Wagner, G. P. 2007. The developmental genetics of homology. *Nature Reviews Genetics*, 8, 473–479.

Wang, L., Israel, J. W., Edgar, A. et al. 2020. Genetic basis for divergence in developmental gene expression in two closely related sea urchins. *Nature Ecology & Evolution*, 4, 831–840.

Wang, X. & Sommer, R. J. 2011. Antagonism of LIN-17/Frizzled and LIN-18/Ryk in nematode vulva induction reveals evolutionary alterations in core developmental pathways. *PLoS Biol*, 9, e1001110.

Watson, J. D., Baker, T. A., Bell, S. P. et al. 2008. *Molecular biology of the gene*, San Francisco: Pearson Education.

Wessel, G. M., Kiyomoto, M., Shen, T.-L. & Yajima, M. 2020. Genetic manipulation of the pigment pathway in a sea urchin reveals distinct lineage commitment prior to metamorphosis in the bilateral to radial body plan transition. *Scientific Reports*, 10, 1–10.

Wilt, F., Croker, L., Killian, C. E. & Mcdonald, K. 2008. Role of LSM34/SpSM50 proteins in endoskeletal spicule formation in sea urchin embryos. *Invertebrate Biology*, 127, 452–459.

Wilt, F., Killian, C. E., Croker, L. & Hamilton, P. 2013. SM30 protein function during sea urchin larval spicule formation. *Journal of Structural Biology*, 183, 199–204.

Woodland, W. 1906. Memoirs: Studies in spicule formation: IV.–The scleroblastic development of the spicules in cucumariidæ; with a note relating to the plate-and-anchor spicules of synapta inhærens. *Journal of Cell Science*, 2, 533–559.

Woodland, W. 1907a. Memoirs: Studies in spicule formation: V.–The sclero-blastic development of the spicules in ophiuroidea and echinoidea, and in the genera antedon and synapta. *Journal of Cell Science*, 2, 31–44.

Woodland, W. 1907b. Memoirs: Studies in spicule formation: VII.–The scler-oblastic development of the plate-and-anchor spicules of synapta, and of the wheel spicules of the auricularia larva. *Journal of Cell Science*, 2, 483–510.

Wörheide, G., Dohrmann, M. & Yang, Q. 2016. Molecular paleobiology—progress and perspectives. *Palaeoworld*, 25, 138–148.

Wray, G. A. 1992. The evolution of larval morphology during the post-Paleozoic radiation of echinoids. *Paleobiology*, 18, 258–287.

Wray, G. A., Hahn, M. W., Abouheif, E. et al. 2003. The evolution of transcriptional regulation in eukaryotes. *Molecular Biology and Evolution*, 20, 1377–1419.

Wray, G. A. & Lowe, C. J. 2000. Developmental regulatory genes and echino-derm evolution. *Systematic Biology*, 49, 28–51.

Wray, G. A. & Mcclay, D. R. 1988. The origin of spicule-forming cells in a "primitive" sea urchin (*Eucidaris tribuloides*) which appears to lack pri-mary mesenchyme cells. *Development*, 103, 305–315.

Wright, D. F. 2017. Phenotypic innovation and adaptive constraints in the evolutionary radiation of Palaeozoic crinoids. *Scientific Reports*, 7, 1–10.

Wu, S. & Mcclay, D. R. 2007. The Snail repressor is required for PMC ingression in the sea urchin embryo. *Development*, 134, 1061–1070.

Wu, S. Y., Ferkowicz, M. & Mcclay, D. R. 2007. Ingression of primary mesen-chyme cells of the sea urchin embryo: A precisely timed epithelial mesen-chymal transition. *Birth Defects Research Part C: Embryo Today: Reviews*, 81, 241–252.

Yaguchi, S., Yaguchi, J., Suzuki, H. et al. 2020. Establishment of homozygous knock-out sea urchins. *Current Biology*, 30, R427–R429.

Yamazaki, A., Furuzawa, Y. & Yamaguchi, M. 2010. Conserved early expres-sion patterns of micromere specification genes in two echinoid species belonging to the orders clypeasteroida and echinoida. *Developmental Dynamics*, 239, 3391–3403.

Yamazaki, A., Kidachi, Y., Yamaguchi, M. & Minokawa, T. 2014. Larval mesenchyme cell specification in the primitive echinoid occurs independ-ently of the double-negative gate. *Development*, 141, 2669–2679.

Yamazaki, A. & Minokawa, T. 2015. Expession patterns of mesenchyme specification genes in two distantly related echinoids, Glyptocidaris crenu-laris and Echinocardium cordatum. *Gene Expression Patterns*, 17, 87–97.

Yamazaki, A., Morino, Y., Urata, M. et al. 2020. Pmar1/phb homeobox genes and the evolution of the double-negative gate for endomesoderm specifica-tion in echinoderms. *Development*, 147, dev182139.

Zamora, S. & Rahman, I. A. 2014. Deciphering the early evolution of echino-derms with Cambrian fossils. *Palaeontology*, 57, 1105–1119.

Zhang, X., Sun, L., Yuan, J. et al. 2017. The sea cucumber genome provides insights into morphological evolution and visceral regeneration. *PLoS Biology*, 15, e2003790.

Acknowledgments

I would like to thank Dave Bottjer, who first gave me a copy of "Paleogenomics of Echinoderms" in 2011, while I was an undergraduate visiting his lab, and for giving me the opportunity to explore the world of molecular biology and GRNs during my subsequent Ph.D. thesis work under his tutelage. I would also like to thank Colin Sumrall for the invitation to submit this Element and participate in the accompanying short course. This manuscript benefitted from conversations with E. Erkenbrack, P. Oliveri, L. Piovani, and I. Rahman, and from comments from I. Rahman, N. Wood and E. Petsios. I would also like to acknowledge a Leverhulme Trust Early Career Fellowship for funding and the insightful comments of two reviewers.

Cambridge Elements ☰

Elements of Paleontology

Editor-in-Chief

Colin D. Sumrall
University of Tennessee

About the Series

The Elements of Paleontology series is a publishing collaboration between the Paleontological Society and Cambridge University Press. The series covers the full spectrum of topics in paleontology and paleobiology, and related topics in the Earth and life sciences of interest to students and researchers of paleontology.

The Paleontological Society is an international nonprofit organization devoted exclusively to the science of paleontology: invertebrate and vertebrate paleontology, micropaleontology, and paleobotany. The Society's mission is to advance the study of the fossil record through scientific research, education, and advocacy. Its vision is to be a leading global advocate for understanding life's history and evolution. The Society has several membership categories, including regular, amateur/avocational, student, and retired. Members, representing some forty countries, include professional paleontologists, academicians, science editors, Earth science teachers, museum specialists, undergraduate and graduate students, postdoctoral scholars, and amateur/avocational paleontologists.

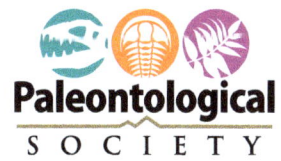

Paleontological
S O C I E T Y

Elements in the Series

A full series listing is available at: www.cambridge.org/EPLY

12
301

Lightning Source UK Ltd.
Milton Keynes UK
UKHW021330091222
413518UK00020B/276